The Cosmos Economy

Jack Gregg

The Cosmos Economy

The Industrialization of Space

Copernicus Books is a brand of Springer

Jack Gregg
Whittier, CA, USA

ISBN 978-3-030-62568-9   ISBN 978-3-030-62569-6 (eBook)
https://doi.org/10.1007/978-3-030-62569-6

© The Editor(s) (if applicable) and The Author(s), under exclusive license to Springer Nature Switzerland AG 2021
This work is subject to copyright. All rights are reserved by the Publisher, whether the whole or part of the material is concerned, specifically the rights of translation, reprinting, reuse of illustrations, recitation, broadcasting, reproduction on microfilms or in any other physical way, and transmission or information storage and retrieval, electronic adaptation, computer software, or by similar or dissimilar methodology now known or hereafter developed.
The use of general descriptive names, registered names, trademarks, service marks, etc. in this publication does not imply, even in the absence of a specific statement, that such names are exempt from the relevant protective laws and regulations and therefore free for general use.
The publisher, the authors, and the editors are safe to assume that the advice and information in this book are believed to be true and accurate at the date of publication. Neither the publisher nor the authors or the editors give a warranty, express or implied, with respect to the material contained herein or for any errors or omissions that may have been made. The publisher remains neutral with regard to jurisdictional claims in published maps and institutional affiliations.

Copernicus is part of Springer, an imprint published by Springer Nature
The registered company is Springer Nature Switzerland AG
The registered company address is: Gewerbestrasse 11, 6330 Cham, Switzerland

This book is dedicated to family, friends, and colleagues who put up with my bad jokes and my contrarian point of view. Thanks!

Acknowledgments

I confess. I did not do this project alone. I had lots of help.

I had help from my family. My wonderful wife Chihiro, my brilliant son Thomas, my loyal sister Jan, my parents Sandy and Bee, and my forebears who got fed up with centuries of hate and violence in Europe and reasoned that if they were destined to be poor and destitute then why not be poor and destitute in America.

I had help from my friends still alive and a few now dead. Especially, my oldest friend Dr. Alex "Buzz" Lucas because he did not believe in bullshit. Also, my dear friends and colleagues Dr. Dave Stewart, Greg Miller, Dr. Ray Haynes, Dr. Mark Allen, Dr. Lesli Handmacher, Sherry Benjamins, Mark Ernst, Robin McNatt, Randy Williams, Rich Rodner, Katherine Beer, Morris Schneider, Anita Stadler, Dan McKinney, Lev Mckinney, and Jeanne Mckinney.

I had extraordinary help from Prof. David Stewart who encouraged my project and, by asking that I present this topic to his business classes, helped me to more acutely focus on the key elements of the narrative.

I had help from my time spent at the University of Chicago, Syracuse University, the University of Miami, the University of California Irvine, Pepperdine University, California State University Long Beach, the University of California Riverside, Loyola Marymount University, and Whittier College, where I was first a student, next an administrator, and finally an adjunct professor of Leadership, Marketing, and Organizational Management.

I had help from my teachers, students, and colleagues. Dr. Vance Caesar, Dr. David Stewart, Dr. Lana Nino, Dr. Farzin Magidi, Dr. Dennis Aigner among others. Students Ana Gimber and Ty Lopez helped design and launch my web page. Special thanks to Dr. Harris, my pediatrician, who was there from the beginning.

I had help from my book coach, Greg Miller, who helped me find my footing as my writing guide and then became a trusted friend along the way. I had help from my professional coach, Vance Caesar, who helped me create my vision. My gratitude to Nitza Jones-Sepulveda, my editor who said "yes" when others could not see my vision. And McMillian/Copernicus press who will hopefully make a fortune from this book.

I had help from generations of serious scholars, rigorous researchers, and dynamic writers referenced throughout the book.

I had help from the many space professionals and thought leaders who graciously agreed to be interviewed for this project, most notably Shelli Brunswick, Kevin O'Connell, Gwynne Shotwell, Dr. Andy Aldrin, Dylan Taylor, Meagan Crawford, Jeff Matthews, Prof. Lisa Yaszek, Andrea Seastrand, Dr. Jim Keravala, Mandy Vaughn, Dr. Matt Weinzeirl, Rod Pyle, Mat Kaplan, Ivan Rosenberg, Will Pomerantz, Masayasu Ishida, Dr. Greg Autry, Dr. Angela Diaz, Dr. Alex MacDonald, Dick Croxall, Jack Goldberg, Joe Donnelly, and the many others who I have met and talked with on this journey and who have contributed their thoughts and perspectives about the new space economy.

All the other writers and thinkers about space have helped me to bring the space economy out of the shadows of fantasy into the bright light of inevitable certainty.

I had special help from most of my ex-bosses whose stunning lack of vision and leadership skills collectively ignited my desire to leave their stultifying company and escape their insipid dictums to pursue my own vision of success.

And I had incalculable support from my many cats, past and present, who over the years have patiently listened to my rants and ramblings then graciously purred in obsequious agreement. Especially my current coyote survivors, Tora and Stinky.

Praise for The Cosmos Economy

Timely... informative... engaging. Jack Gregg's new book, *The Cosmos Economy*, is a welcome addition to the very short list of books on this topic. *The Cosmos Economy* is both an approachable read and a deep dive on a fast-moving, sometimes complex topic, the NewSpace economy. While we all see the tip of the iceberg in the news media about the activities of SpaceX, Blue Origin, and others, Gregg's book gives us a behind the scenes look at the financial machinery that will drive this exciting new industry.

—Rod Pyle, Author of *Space 2.0*
and Editor-in-Chief of *Ad Astra* magazine

Markets have always had boundaries—until now! *The Cosmos Economy* shares why this truly global and multidisciplinary marketplace will transform how everyone works and lives in the twenty-first century and beyond. Jack Gregg offers the primer to understanding how commerce, investment, and entrepreneurship will be changed by the final frontier.

—Shelli Brunswick, COO, The Space Foundation

If you have heard about the recent revolution in the space sector but want to know more—about the path forward, the implications for life on Earth, and even your role in it—this book is an excellent place to start. Gregg brings both analysis and creativity to his subject, providing a roadmap for the development of the space economy and answers to some of the most-asked questions from skeptics and enthusiasts alike. You will be inspired by the future he projects, and you will be able to use the solid foundation of facts and information he provides to spread that inspiration to others.

—Prof. Matthew Weinzeirl, the Joseph and Jacqueline Elbling Professor of Business Administration at Harvard Business School. Dr. Weinzeirl is a noted author of articles on space economics.

Dr. Jack Gregg has made a substantial contribution to our thinking about the space economy with this volume. He nicely captures the rapidly changing roles of government and industry, including how commercial business practices are increasing the speed and efficiency of new space missions. His focus on the lessons from other emerging markets and the growing trade in space services provides important pointers for entrepreneurs, financial institutions, educators, and government regulators as they enable the space economy. This book is an important volume for everyone concerned with the future of space exploration and space commerce, including the many opportunities they will create on Earth.

—Kevin M. O'Connell, Former Director, Office of Space Commerce, U.S. Department of Commerce

The real Space Age has finally arrived. That's the message of *The Cosmos Economy*, a deeply researched, highly readable and very engaging look at the business promise of the final frontier. The first sixty-plus years of space development were dominated by governments seeking prestige, security, and science. Jack Gregg shows us that the momentum has now been transferred to companies like SpaceX that base their entire existence on what can and will be done beyond Earth's atmosphere, as well as to other, long-established corporations that recognize the nearly limitless potential that awaits. This is a book about opportunities. It's as fine a guide to these as I have ever seen.

—Mat Kaplan, Host and Producer, The Planetary Society/Planetary Radio

Contents

1 A New Destiny   1

Part I   The Business Case for Space

2 The Cosmos Economy   17

3 Blue Oceans and Greenfields in Space   25

4 The Importance of Frontiers   33

5 Coming into the Cosmos   43

6 No Country for Earthmen   53

7 Turning over [Space] Rocks …   61

8 Planning on Purpose   69

Part II   Coming into the Cosmos

9 Settlement Communities   81

10	Forever Frontiers	89
11	Pushback and Challenges	97
12	Visions of Space	103
13	The Space Economy Is Already Here	117
14	The New Space Merchants	127

Part III Heavenly Markets: Getting Extraterrestrial

15	The Industrialization of Space	133
16	Space Biz	135
17	Space Mining	145
18	Space Manufacturing	153
19	No Country for Earth Men	159
20	Colonies, Outposts, Settlements, and Stations	167
21	Don't Look Back	179

Part IV The Rise of Power, Politics, and Policy in Space

22	The New 49ers Rush to Space	187
23	A Competitive Solar System	193
24	The Rules of the Game	201

| 25 | Trouble in Paradise | 209 |

Part V The New Space Economy: Setting the Stage for the Industrialization of the Solar System

26	Industrial Space	215
27	Business Models	223
28	Diffusion of Innovations: The Five-Phase Adoption Model	237
29	Connecting the Dots	257

Part VI Lessons Learned on the Way to the Cosmos Economy

30	A New Generation of Pilgrims	263
31	Questions and Answers	269
32	Lessons Learned	277
33	How to Be Part of the New Cosmos Economy	283

Appendix 1: Research Findings	287
Appendix 2: Recommendations for Future Research	295
Appendix 3: Self-Quiz About Space: How Do you Measure Up?	299
Index	301
About the Author	313

List of Figures

Fig. 4.1   Sectors of the Cosmos Economy. Traditional space activities include (1) military interests by sovereign nation states, (2) scientific exploration such as space telescopes and probes to celestial bodies in our Solar System, (3) civil space activities such as GPS and weather tracking, and (4) commercial space ventures currently in LEO (low Earth orbit) 41

Fig. 20.1 Space Communities. A comparison of several space community formats measured by independent economic sustainability and the degree of relative autonomy............................ 172

Fig. 23.1 This chart plots the dynamic growth of the 5-phase adoption model over time. Superimposed on the 5 phases timeline is an S-Curve illustrating how the space economy will adopt from each phase to the next. Labels of the five phases are adapted from Everett Rogers (1962, 2003)... 198

Fig. 27.1 Think of the area under the curve as representing the potential volume of demand for your product. The obvious place to be is where demand is clearly greatest; directly in the middle of the distribution curve (location A). The thin limited demand off to the right—the long tail of the classic "bell" curve—represents a unique pricing opportunity (location B). This is because the firm can premium price (charge a LOT more for the product). This high-margin solution circumvents the low-margin problem of targeting the middle of the distribution curve where other competitors congregate.. 234

List of Figures

Fig. 28.1 The five phases are not evenly adopted. The first three phases, representing higher risk, are equal to the last two phases when there is higher confidence. The steepest growth rate occurs in phase 3 between the two inflection points represented by the arrows on the figure 240

Fig. 28.2 Phase 1: Innovator Phase (Frontier Phase). Earth is central hub of Frontier/Innovator Market Economy. Economic activity is defined and controlled by Earth-based activities such as commercial investment, technology, and rudimentary mining and space-based manufacturing. Destinations X, Y, and Z indicate the possibility of various industrial outposts, settlements and commercial activities including cislunar, asteroids, planets, or deep space locations. This phase will see investments in primary infrastructure. Space policy agreements will begin to be challenged as industrial momentum increases... 244

Fig. 28.3 Phase 2: Early Adopters. Resolution of the "down-mass" problem will determine direction and focus of this phase of development. If down-mass technologies are developed, then Earth-based demand will drive the space economy. If this problem is not resolved, then space-based demand will define the cosmos economy. Space-driven economy will set a tone of greater independence from Earth and focus growth on other markets in the solar system. Frontier to Early Adopter transition. Trade is still highly dependent upon Earth as a market partner. Note: the "?" indicates the unresolved question about the commercial ability to ship large payloads of goods from space producers to Earth markets: The down-mass problem 247

Fig. 28.4 Phase 3: Early Mainstream. Trade among established space centers indicates increasing independence from Earth. Imposition of regulations and standardized accountability metrics. New outposts, trade hubs, far-flung communities, and urban centers. Increase of consumer-related goods and services 249

Fig. 28.5 Phase 4: Late Mainstream. Expansion to Mature Market transition. Trade is no longer reliant upon Earth-based markets (represented by dotted line encircling Earth). Greater participation of consumer-oriented manufacturing... 252

Fig. 28.6 Phase 5: Lagging Adopters (AKA: Maturity Phase). Mature/Independent Market Economy. Closed loop: Self-sustaining demand and supply completely in space. Establishment of new commercial hubs and communities. Growth will come from pushing farther out in the solar system, thus creating an endless succession of new frontiers... 254

List of Tables

Table 4.1	This summary assessment of old and new assumptions about a frontier economy shows that while some elements persist (e.g., cost of transport, exploitation of natural resources) the impact of technology and big corporations will have a greater effect launching the space frontier economy	39
Table 8.1	The new space economy will evolve in five overlapping phases of adoption. Each phase represents a successive era of economic activity that will benefit from prior ventures. Different business models, industries, and enterprises will align fittingly for each phase of adoption. Date ranges are notional	76
Table 26.1	In this table the phases of adoption are shown sequentially. However, when we consider the immense domain of space it is highly likely that different phases may be adopted faster than others or may even progress in a non-linear way. Anticipated date ranges are approximate and intentionally overlap.	221
Table 27.1	This table shows the relationship of a few selected common business models along with (a) likely optimal adoption phase(s), and with (b) a set of commercial entities likely to employ the business model for each phase(s) of adoption in the space-based economy	226
Table 28.1	The five phases of adoption are shown with approximate dates, likely business models that best fit each phase, and industries most likely to prosper in a given phase	238
Table 28.2	Phase 1 Synopsis. Please note that all references to companies, organizations, and sovereign nations are for purposes of illustration only	241

Table 28.3	Phase 2 Synopsis. This table shows the relationship between phase 1 and phase 2. Phase 2 is strongly dependent upon the degree of success of the first phase in establishing a successful foundation. With a broad spectrum of different business sectors established in phase 2 comes a mix of business models and organizations eager to enter and adopt the new environment	245
Table 28.4	Phase 3 Synopsis represents the early mainstream adoption. In addition to an expansion of industry sectors and markets this phase is noteworthy because of the development of intra-space trade as the major focus of industrial space economic expansion	248
Table 28.5	Phase 4 Synopsis. With the coming of investors in phase 4, adoption is estimated to reach roughly 84% saturation. This heralds a mature market with sellers competing more on price in lieu of such intangibles as quality and service. Earth is no longer an integral component	251
Table 28.6	Phase 5 Synopsis. This is the full five-phase model. Dates indicated at the top of each column are approximations to suggest an overlapping sequence of progression. The nations, organizations, and firms are represented as likely examples only	254
Table 29.1	The complete adoption model synopsis. This figure shows the combined relationships between the five phases, likely business models, industries, and markets, along with the types of firms that will conduct business in various space settings. The dotted line S-Curve superimposed on the figure shows the cumulative degree of adoption from early frontier through late adopter phases	258
Table 31.1	This comparison demonstrates similarities of emerging markets on Earth and the Phase-1 space frontier. There are also distinct dissimilarities such as rents, labor costs, the impact on indigenous peoples (on Earth), and the scope of competition especially at initial stages of development	270

Introduction

If you are looking to read about space aliens secretly working at Walmart on the night shift, the mathematical proof for traveling faster than the speed of light, secrets revealed about Area 51, structural diagrams of the Death Star, or tips for packing your knapsack for time travel, then you have come to the wrong place. This is a book about how commercial enterprise will ignite the growth and development of our solar system and spur human settlement in space.

This is about the *business* of space settlement.

The volatile risks of investing in the new space frontier can go either way and can make winners or losers out of investors. Unlimited raw resources offer an exciting enticement of unlimited wealth, but scant infrastructure coupled with the lack of a consumer base is a toxic cocktail that threatens dramatic losses for investors. Given the unpredictable nature of space, why is the new space economy projected to be $3 trillion by mid-century?

This book explores how the promise of profit and power will likely motivate adventurers and entrepreneurs to overcome the life-threatening hardships of space and create a commercial platform for human settlement in our solar system.

This is written for individuals and organizations curious about the prospect of conducting business in the emerging space sector. When I started this project, my purpose was to: (a) provide an entertaining and informative overview of the space settlement economy for the general business audience, (b) create awareness of the emerging commercial space economy, (c) start a rigorous conversation about what it will take to launch a space-based enterprise, and (d) propose a strategic roadmap (vision) for space that is easily understood (i.e., non-technical) by the average business reader, career explorer, entrepreneur, or space enthusiast. The text purposely avoids technical jargon but recognizes that innovative technology will enable human settlement and commercial enterprise.

Right now, there are planners, doers, and dreamers who are thinking about space in a very down-to-Earth way. These space enthusiasts and activists may be an engineer who will design an elegant solution to a difficult technical problem, an entrepreneur who will not rest until she makes her space vision a reality, the corporate strategist who recognizes the rich potential of developing new products and new markets in space, the VC seeking new opportunities to fund ventures at different stages of the space sector's development, and the academic researcher who seeks to learn from this new emergent economy. Also engaged in this discussion is the student and the careerist, dissatisfied with today's diminishing career options, in search of something new and challenging for their life's work. Then, of course, there is everyone else who will be affected by the exploding growth of the new space economy.

Most books about space tend to focus singly on specific aspects of space like astronomy, hard science, rocketry, technology, historical accounts of space missions, or profiles of space celebrities. Likewise, books about business tend to keep narrow focus on leadership, management practice, investment, motivation, or a host of other prescriptive solutions to the challenge of managing a business enterprise. This book is different. It is about how space settlement will be enabled by commercial business activities. It further presents a research-based roadmap for establishing a mature economic presence in our solar system—a process that is already under way and is rapidly gaining international momentum. This hybrid focus of business and space aspires to serve those individuals and organizations who see space for what it is—the next frontier for business supported by expansive human settlement beyond Earth.

This is neither a science fiction fantasy, a turgid technical treatise, nor an arcane academic research paper. Still, there is much here for Sci-Fi enthusiasts, techies/ engineers, and academics who have passionate interests in all things space and futuristic. This interest is enhanced via some major themes in the book such as robotics, space-based manufacturing, the role of humans in space, emergent political/social/economic conflicts, and detailed opportunities for entrepreneurial investment.

This book is designed to appeal to: (1) a general reader who is fascinated with space exploration via their association with popular science publications, space-fi books, films and TV, or the interest created by a continuous barrage of space-related news stories in the popular media, (2) a business-minded reader who wishes to learn about the economic potential of developing space for commercial purposes so they can connect their current or potential business to the new space market, (3) an academic searching for information about today's space entrepreneurs for classroom case discussions and for their own independent research opportunities, (4) a financial investor looking to capitalize on the coming opportunities in the emerging markets of industrialized space, (5) a space policy regulator who recognizes that as space becomes a mature market economy they will have a greater role to play, (6) an employee of an aerospace firm who is already deeply engaged in developing the space-based economy, (7) a leader or manager of a multinational firm that is not currently branded as space enterprise and who is seeking new opportunities to pivot and adapt to the emerging space-based economy, (8) a representative of a sovereign nation who is committed to developing off-planet activities for a broad range of economic, political, military, and ideological reasons, and (9) a careerist, entry- to mid-level, who seeks information and validation about how to best position themselves for a rewarding and adventuresome career in the space business sector.

If you fall into one (or some) of these categories or if you simply have an interest in man's future in space, then you are in the right place. Read on. If you are thinking about a career in commercial space, or thinking about investing in the burgeoning commercial space sector, or considering building a business venture that will benefit from the new space economy, or simply want to get a better understanding of civilization's next great leap forward into space, then read on. If you picked up this book by mistake or were given it as a gift by your least favorite co-worker at a company potluck gift exchange, then fear not. You are still in the right place. Read on.

There is a strong chance that you have already developed a set of notions and assumptions about space. There is a stronger chance that you have not given much thought to how space will affect your life or your kids' lives. It is not news that today's professional will change industries multiple times in her professional life—will the new space economy be a career option on your horizon? The incessant barrage of news about reusable rockets, internet from satellites, space tourists paying big bucks for the joy ride of a lifetime (souvenir space suit included), and the buzz about big corporations and even bigger billionaires investing in space probably was not on your radar screen a decade ago. But it is now.

It is no secret that we are on the threshold of an exciting new era of human progress. Amidst all the optimism and utopian expectations about life in space, a bit of healthy skepticism is in order. Establishing the new space frontier will probably do nothing to mitigate Earth's historic problems of famine, war, poverty, fuzzy thinking, hatred of the others who live among us, and social injustice.

The Earth Is a Harsh Master

Life on our planet is not easy. Most plants and animals spend their brief existence in a constant fight for survival and reproduction. Death is not only the inevitable outcome at the end of the struggle but, sadly, the most important provider of nutrients for future generations up and down the food chain. As for humans, at the height of our collective technology now in the early twenty-first century, most of our species live in squalor and poverty. Political trends of exclusion rather than inclusiveness have made tribal enemies of those "others" who threaten us, just because. Global climate change threatens the prospect of famines and mass migrations of people seeking a chance for a better life away from destitution and oppressive political policies. We have not yet created a planet that is a home for all of us. One can say the march of civilization is a retreat from fear, the fear of starvation, the fear of enemies, the fear of disease, the fear of living in a naturally hostile world.

When space settlement is discussed, it is common for a voice or two to remind us that living in space will be hard, that resources for our survival will have to be wrung from the local environment, that there are no guarantees of survival, that people living in space will live in fear of a naturally hostile world. But that is what our species has been doing for countless thousands of years as we have tried to tame hunger with agriculture, make our lives safer with medical science, control our world with technology, and create a more meaningful life through art.

If we have been able to negotiate a state of détente with our harsh master, Earth, then surely there is hope for us to take the lessons we have learned from our new home in space.

There are questions about who will benefit most from a new industrialized space. How will new wealth from space industries impact old wealth back on Earth? Will the proliferation of outposts and settlements in space disrupt the current political balance of nations back on Earth? And if centers of political and corporate power are repositioned, then what will be the social ramifications for those who are disenfranchised by this shift; will the initiative to settle space cause a greater gap between the have-nots from the haves?

Tendencies to tribalism, what has lately come to be called "identity politics," may not soon disappear. As Seth Godin explains in *Tribes* (2008) people will still claim affinity with others of their religion, social class, geographic origin, gender preference, political affiliation, zip code, shoe size, etc. Some will continue to align with others just like themselves. Some will define themselves in opposition to the "other." What remains to be seen is if the space frontier will provide opportunities for pioneering individuals to rebrand themselves, to start afresh, in the same way that immigrants have always created new identities in their newly adopted homeland. Will the future generation of space settlers prefer to be recognized as Martians? And will the new Martian tribalists prefer to think of themselves as superior to people on Earth or to settlers on some other chunk of real estate in our solar system? Or just different folks?

If so, then they will merely be swapping an old identity label for a new one and perpetuate the old trope of claiming distinction by being different (i.e., being better) than the "other" instead of removing those old tribal barriers that kept us apart and at odds with one another for thousands of years. Being a member of a team means that non-members are different, are competitors for limited resources, are potential enemies, are convenient targets of ridicule and, often enough, subjects of persecution. Bringing human civilization to space brings no guarantee that this ageless tendency to chauvinism is likely to go away. Our sophisticated technologies may evolve, but there is no guarantee that our primitive insecurities will change. In this regard, and others, the promise of a utopian era in space is a likely false idol.

If settling space does not produce the magic cure that will signal a new age of human maturity, then what hope is there for one of the oldest human transactions: trade? At the heart of a sustainable long-term business relationship is trust. But human nature, such as it is, tends to distrust strangers and can impede new business opportunities. Will future Mars folks distrust future Earth folks just because they have a different planet identity? Probably.

Space, like other brave ventures, is a subtle combination of *going to* a new place and *leaving from* an old place. There is resistance and attraction at both ends of this push/pull emotional tug of war. Starting a college career is as much about *leaving* childhood as it is about *going to* future adulthood. Changing a career is an action of abandoning the limits and constraints of the old job as well as adopting the challenges of doing something new. Columbus' cautious crew were likely as desperate to leave Spain's brutal inquisition as they were eager to share in the rewards of forging a new route to India. There are doubts and regrets as well as optimism and hope when looking backward or forward in any transition. Major reasons to leave Earth are easy to catalog. Human population is growing at a rapid rate (over 10 billion people by mid-century). More people living on a planet made ever smaller by an inability to provide enough food or shelter is a scary prospect—and a good reason to abandon ship. Here are a few more reasons:

- Global resources are being depleted—even with the best intentions we will not be able to match expected demand to support everyone in only a few decades. Can space provide a solution?
- The ability to feed, house, and govern our growing population is declining—efforts to remedy this problem (e.g., clearing the Brazilian rainforest to create more arable land, developing more potent fertilizers to increase crop yields) are not sustainable in the long run and are facing a flattened curve. Can the new frontiers of space solve this dilemma?
- Global forces such as the threat of war, deadly pandemics like the Covid-19, the impact of climate change, the clash of cultures, and conflicting religious faiths add a general sense of tension and anxiety to our lives. Will space offer a sounder sanctuary?

- Jobs at all levels are being made vulnerable to robotic replacement thus causing uncertainty about what life will be like in the future. This is not just a "blue collar" concern. Advances in AI have advanced rapidly through all organizational ranks. Is space immune from the robot revolution?
- The trend to mass conformity by big brother totalitarian states like China is a scary prospect for liberal-minded western citizens. Will new settlements in space offer an escape from dictatorial encroachment on individual autonomy?

1

A New Destiny

When I learned about Manifest Destiny in my high school history class, I took away the patriotic notion that America was a promised land. A place, as the phrase implied, where there was a grand purpose of divine origin that would yield success and prosperity to those who took bold actions, who pursued a vision of economic growth, and who would be rewarded for the risks they took. Simply put, I was taught that the acquisition of new territories was a natural progression of the development of the American experiment in the New World. I thought little about the indigenous peoples who already lived in the frontier expanse or of the impact on the pristine environment by unfettered industrialists, who were the primary beneficiaries of the annexation of new lands and resources. I was taught, incorrectly, that manifest destiny was a sort of nationalist religion, a rallying focus that celebrated the territorial expansionism of the nineteenth century. Something that promoted ambition as a social value to the exclusion of others. I was not taught that by the end of the nineteenth century, the idea of a manifest destiny became a contentious social, political, economic, and even legal issue. I did not know (or do not recall considering) that there was vocal opposition to the idea of a manifest destiny. I did not know that the author of the original phrase reportedly meant it as a sarcastic snipe, a criticism of those who would bulldoze their way across the frontier with the self-serving intent to acquire and exploit vast new land tracts for their own private control and profit.

So, when I interviewed space industry leaders for this book, I naturally expected a degree of bias because, after all, they lived and breathed the passionate dream of space settlement, they were committed to creating a part of the new space economy. But I was surprised to hear many impose caution on their collective dream. Caution about doing no harm to virgin celestial asteroids, moons, and planets. Caution about making sure that developing space was not seen as a zero-sum game where taking care of ourselves here on Earth would be diminished in proportion to resources applied to developing space. Caution about the potential heavy hand of some nations and corporations to dominate and control the resources of the solar system at the expense of other, less powerful, firms and states. And caution about the simplistic and naive dream of future mankind living in space settlements, like the Jetsons, in a fanciful perfect space utopia. They cautioned me to consider the development of space with eyes wide open and fantasies of a future Shangri-La held firmly in check.

To be sure, there are and will be those who see the global interest in the commercial settlement of space as something far less than beneficial to the human narrative. There will always be contrary perspectives on civilization's important milestones. The skeptics and opponents of space development may be compared to the Luddites who opposed the mechanization of the English textile mills and the automation of jobs in the mills just as the industrial revolution shifted into high gear in the 1820s. Every age has seen a version of Luddites reject new forms of technology—workers attempted to sabotage and reject office and manufacturing computerization in the latter half of the twentieth century, for example.

Managing Change

New methods and new markets will undoubtedly threaten the status quo. The idea of something big, consequential and new like the impact of commercial space development on traditional manufacturing, transportation, commerce, finance, manufacturing, and other economic activities may threaten the traditional way of doing business. But is that a bad thing?

New trade agreements opened markets and cultural influence between nineteenth century Meiji Japan and the west via "gunboat diplomacy." Colonial expansion brought new goods to Spain, Portugal, Britain, France, Germany, the Netherlands, and to other nations seeking to leverage cheaper labor along with untapped agriculture and mineral resources in far-off developing regions. As Krasavina describes in *Colonial Trade* (2010), it was messy, to be sure. It was often cruel and exploitive. It was also the cradle of new independent nations in the twentieth century who aspired to compete in the global economy on their own terms. The new space industrial age will certainly invite comment from those who reject the development of space because they may view space as a threat to their current way of life, something to be feared instead of embraced. Naysayers to technology advancements at the dawn of the industrial revolution, the Luddites famously tried to hold back the tide of industrial progress by attacking the new machines. Andrews (*Who Were the Luddites?*, 2019) observed that there will always be Luddites among us who may have legitimate concerns about the impact of new technologies and change—human nature and greed being what it is—but this time around there are no indigenous peoples to exploit or kill, no rival spacefaring nations to confront or fight, no trade routes or manufacturing centers to protect. Not yet, anyway.

The Challenge of Writing About the Future

Exploring the future of anything is a challenge. How can you get the facts right if the facts are not yet facts? If the topic is the future of human space settlement, then the task is especially problematic. I found myself writing from the perspective of an outsider looking in on a topic through a very cloudy window. I knew the experts had opinions that they passionately held, and I recognized that this was both good and bad. For a fluid and evolving topic like the industrialization of our solar system, experts come in several varieties: Quants and poets.

Quants are experts from the technical and engineering disciplines who tend to be highly analytical and rigorous in their thoughts about how the physical world can be manipulated to achieve this or that escape velocity, criteria for human habitat in cislunar space, in situ resource utilization (ISRU) on Mars or on Jupiter's moon, Io. Quants are indispensable in studying space. They do the heavy lifting of analyzing, designing, developing, implementing, and evaluating the technologies that make space a viable subject. They often see space as a vast lab bench ripe for innovating new hardware or software solutions.

Poets, on the other hand, tend to be more focused on the general schema of space. Theirs is the realm of corporate strategists, entrepreneurs, venture capital fund managers, space tour operators, and others who see the prospect of settling space through a long-term lens, as something much more than technological achievement. Poets see space settlement in terms of social, political, environmental, legal, and economic impact. Their focus is on the human imperative of space. They see the potential of building new space-based markets from a business-minded point of view. To be sure no one in these simplified categories is either fully a Poet or a Quant. People in each category have a bit of the other's talents and skills. In fact, the best experts tend to have a blend of right- and left-brain traits.

This central balance of perspective and capabilities between poets and quants was dramatically framed in Kurt Vonnegut's first novel, *Player Piano* (Vonnegut 1980/1952). Vonnegut's wonderful satire (based on his early corporate experience with General Electric) paints the future in bleak dystopian strokes where the society is automated, controlled by huge corporations, and led by over-educated engineer and manager elites. The engineers are there to keep the machinery running and the managers are responsible for the health of the bureaucracy. Only those with a Ph.D. are employed (even real estate agents have doctorate degrees) while everyone else lives on a state-provided stipend, a universal basic income (a foreshadow of today's current debate), to keep money flowing and the economy alive.

As I began to explore the future of commercial space activities and as I attempted to suss out the enabling trends and the inhibiting speedbumps of the new space economy, I believed that I could bring a fresh perspective to the topic about what was to come. I felt that I could be the helpful skeptic in the conversation able to challenge and explore the notion of a new space economy from a business mindset instead of from the traditional technical viewpoint that has defined human space exploration for well over half a century. The focus of this book is the achievement of that goal.

It is easy to research the past. A book about the early years of space can easily rely upon mountains of readily available documentation, analytic reports, academic studies, individual memoires, and interviews with famous folks who were central to historical events. But there are scant resources that catalog the future. The problem is not only that the history of the future is not yet writ, but as a researcher, I confess that the challenge is, as Defense Secretary Donald Rumsfeld famously said, "You don't know what you don't know." (2002) Learning something new is both the fun of the hunt and the frustration of dead-end inquiries. I have interviewed academics, space business professionals, government leaders, and representatives of foreign space agencies. Each has their own notion of the future, a vision, that sometimes follows its own vector and occasionally overlaps with another's idea of what space will be. But the dominant truth is that no one knows what they do not know about the future of space. We will all have to stay tuned.

This topic is a snapshot of a grander work in process: civilizing the solar system. Most of the current commercial space activities center on ventures in Low Earth Orbit (LEO) such as imaging, communication, data analytics, and the launch services that move payloads into space. But today's commercial space activities are decidedly a short-term prospect. My focus is a much longer term. Space settlement, whether on Mars, our Moon, or somewhere cislunar (located in space between the Moon and Earth), is a long-term engagement requiring hefty amounts of capital, political will, and social acceptance. The secret sauce that will add spice to the establishment of space settlements throughout our solar system is the prospect of profit and power for nations and companies willing to make the investments and take the heavy risks.

There are lots of ideas about the economic development of space, but there is no consensus. Over the years, a cadre of thought leaders and business influencers have emerged to define and reflect collective thinking about how space will mature, what business areas will prosper, the role of public and private sectors, the potential market capitalization of space, and the endless catalog of barriers to conducting extraterrestrial business. The challenge is to capture these ideas and distill them into a narrative of likely events to create a portrait of the cosmos economy.

Guessing the Future

Because the future has not arrived yet many of the guesses and conclusions presented in the following pages follow an inductive approach. That is, the outcome or consequence of current government policies, business decisions, or technical developments may occur as we predict, but there may also be another outcome totally unexpected. Following this approach, it would be inaccurate to claim future specifics of space settlement as certain. I prefer to induce that the outcomes presented are at minimum plausible and at maximum highly likely, given current information. The long-term vision of industrialized space may be described as likely or reasonable but cannot be asserted with complete cast-iron certainty. Such are the limits of speculative investigation about the future. Time will be the best judge of the predictions presented here.

There are similar predictive limits when analogies are employed. Especially historical analogies about colonization and civilizing undeveloped frontier territories as I will do later. Comparing the similar properties of two social/historical settings may take us only so far before we must stop and acknowledge the limits of this approach. Just because the settlement of the American frontier in the nineteenth century has some similar traits with the anticipated settlement of space does not mean that the analogy of these compared examples makes one predictive of the other. There are limits because historical circumstances are not alike. Still, there are lessons we can learn by studying examples from our recent past that are likely to apply to the settlement of space.

While the body of substantive research about the new space economy is embryonic, there are research resources that help to provide answers for many of the questions about the cosmos economy. Data exists for those seeking information about investing in space ventures, looking for information about new career opportunities, or planning to enter the space sector with a new business plan. An example of a predictive model to help answer these questions is the five-phase model of adoption, presented in sect. 5, which is designed to parse the different stages of development from the beginnings of a frontier economy through a mature market. Adapted from the work of Dr. Everett Rogers (*Diffusion of Innovations*, 1962, 2003), the model presents a dynamic picture of the likely development phases of the space economy over time. The model, adopted for the developing space economy, describes the economic characteristics for each phase, the types of business models most appropriate for that phase, the industrial sectors likely to thrive, the nature of demand, barriers to entry/exit, the impact of technology, and even the personality profile of a typical entrepreneurial adopter for each phase of space development.

The (*Un*)Willing Suspension of (*Dis*)Belief

Works of speculative non-fiction are about things in the making and not yet fully realized. The best assumptions about the probable, the possible, or the most likely can seem fiction-like; a best guess about the future. Like this book. But if this *were* a work of pure fiction, there would be some notable heroes and villains, dramatic conflicts and confrontations, theatrical tension, moments of high success, and discouraging failure. Maybe even a little love interest thrown in just to keep things interesting. As it turns out all those dramatic fictional elements are here. Except this is not fiction.

As outlined in *Super Structure; The key to unleashing the power of story*, by James Scott Bell (2015), every good story starts with a problem that causes pain and unhappiness to the hero that results in a quest or challenge. Sam Spade's partner gets killed, Dumbo is separated from his mom, Dorothy is threatened with the loss of her dog, Toto. Now we know who the lead character is, their motivating challenge, and what they must do to solve the dilemma by the end of the story. From here it is a matter of how the hero deals with the problem and what they learn about themselves on their journey of self-discovery. Against this backdrop of conflict, an antagonist emerges; The Wicked Witch, Dumbo's cruel circus owner, the gang of killers obsessed with the Maltese Falcon. At the end of all the confrontations, successes, disappointments, betrayals, and conflicts, the hero experiences a profound change. Sam Spade solves the crime and learns about the price of loyalty, Dumbo discovers that a magic feather does not make you fly—it is self-confidence, and Dorothy learns that happiness is not somewhere over the rainbow but right back on the farm with the safety of her family and friends: There is no place like home.

Heroes and heroines, lost opportunities and rewarded risks, competitors and rivals, battles and confrontations, fortunes made, and grub stakes lost. Through it all leaders and visionaries will emerge from the herd and become the next Henry Huntington, Bill Gates, or Leland Stanford to make an indelible mark on the young history of space settlement.

So it is with this narrative tale of space.

The fictional story I would write would be about a visionary with drive and ideas about launching an enterprise in space. She does not know the lay of the land beforehand, but by the end of the story she learns and grows as a person and as a professional through her efforts. Her antagonists are rivals, competitors, and naysayers who tell her the risks outweigh the potential rewards. Her battles are with the challenges of business, the hurdles of operating in space, the uncertainties of the new space economy, and the changing terrain of operating in a fluid and dynamic market.

She risks death in her adventure. Real death because space is a deadly place. Professional death because doing business in a frontier economy is extremely risky to her reputational capital. And the risk of phycological death because she will have to keep her own council on her very isolated and lonely quest.

Other characters in this story would be on-again, off-again love interest in the form of venture capital investors who come courting as well as the major players of the new space industrial complex who are forever on the hunt for a promising start-up to acquire. There may be some passionate institutional flirting but in the end our protagonist remains unseduced by the promise of a big buyout payday with stock options. Or does she?

Antagonists in this non-fiction thriller range from the unpredictable forces of natural events in space, the fragility of over-promised technology performance of critical production components, the unreliable links in her supply chain, the encroaching imposition of powerful nation states and their policies that may alter her business model, or simply the fact that others have figured out a way to get a similar product or service to market quicker, faster, and cheaper.

Of course, every good hero needs a loyal sidekick. Dumbo needed Jiminy Cricket, Dorothy needed the Scarecrow and her dog Toto, and even Sam Spade relied on his faithful secretary, Effie. In this story, our hero is abetted by true believers on her board, advocates from her funders, and even space advocacy organizations and national space agencies who, like a *Cosmos Chamber of Commerce*, play an active role supporting commercial ventures wherever possible.

Since I have not written the story I do not know for sure if she prevails. But I bet that she does. It all depends on her ability to pivot her plan with agility and how she plans her entry and exit strategy. Fortunately for her, she will have a better idea how to do this by the end of this book!

How I Came to Be Passionate About the Coming Space Economy

For over 25 years, I designed and managed graduate level business and management programs (MBAs) and custom Executive Education for top-ranked business schools at highly respected universities. As an adjunct management professor, I taught graduate courses in Leadership, Strategic Marketing, Organizational Behavior, and other business topics to mid- and upper-level executives who were employed at large and small companies. As a consultant and executive coach, I worked to help managers and leaders improve their effectiveness. But it was when I was asked to be the Executive Director of a unique organization supporting the California space industry that I started to learn about the huge economic potential of the space economy. As a result, I became a passionate advocate for future space-based enterprise and how it will ultimately alter the balance of our planet's global economy. During my time as a leader of a space professional organization, I began to develop some nagging questions about the future of commercial enterprise in space. Seeking answers to these and other questions about the future of space led me on a path to write this book. I also wanted to learn about the types of companies that were likely to participate in this new arena, what types of business sectors were likely to thrive, and what business models would generate the greater return on investment. In short order, I had a list of over two dozen research questions built around my central theme of doing business in space. But when I started asking these questions of industry leaders and academics, I was often met with a blank stare and a response like, "Gee, I never thought about that before."

This got me thinking. If space experts were not asking these questions about the future of the space economy, then either (a) I was way ahead of the curve, making this a great topic for research, (b) I was crazy to think about the future space economy and to even ask these questions since apparently no one else was curious about these topics, (c) I just was not asking the right questions, or (d) there was some deep dark secret about the new space economy that no one wanted to talk about. (Hint: there are nuggets of hidden truth in all four of these options, as I came to discover.) These unanswered nagging questions put me on my quest to map the future development and character of the emerging space economy.

What Does "Living and Working in Space" Mean?

Everyone, and no one, is an expert about the future. Strategic analysts, thought leaders, and academics try to suss out what is over the horizon by looking backward to the lessons of history in order to forecast the future based on what has already happened. Corporate business planners spend briefcases of cash on consultants to distill research data and make decisions about where the market is likely to go because they want to be there ready for business when the customers show up. Then there are leaders who act and make the future happen regardless of what the analysis or trends tell them. Their credo is "no guts: no glory" because they believe there can never be results without decisive action.

James Watt launched the industrial revolution when he recognized the industrial potential of Thomas Newcomen's experiments with his primitive steam engine. Others like Thomas Edison, Henry Ford, Orville and Wilbur Wright, Albert Einstein, and two centuries of innovation have transformed the political, economic, and social dynamics of our global civilization from a labor-intensive agrarian economy to a global industrial network of trade. Even Watt could not predict the impact of mechanization on manufacturing and the departure from dependence on cheap human labor. An unanticipated societal benefit of the industrial revolution was aiding the abolition of slavery. As the integration of mechanization spread across various economic sectors and impacted the financial and social structures of production centers, it became clear that it was more economical to employ machines (in agriculture and manufacturing) than to purchase and maintain human slaves. Ethical and moral imperatives aside, it was finally the cold-hearted realization that automated production methods were more uniform, less costly, and more competitive than relying on forced inhuman labor that led to the end of slavery in the industrialized world. Will the unanticipated social benefits from the industrial space economy enable greater equality of race and gender?

These benefits may likely come from advances in technology or creatively tweaking business models to fit the twists and turns of an emerging economic opportunity. Underlying this process of creativity and innovation are the forehead-slapping revelations that encourage abandoning long-held assumptions about the old ways of work and replacing them with new innovative visions of work, production, and trade. These moments of insight and clarity about the future are easier to recognize in hindsight for most of us. But for the true visionaries among us, like those working today to enable tomorrow's robust space-based society, the new developments awaiting us just around the corner are palpable and predictable. Our best hope is to ride along on their coattails as they thread the needle of leveraging emerging technologies, navigate financial risk, and advocate for social acceptance of our future in space.

What Is Next?

The following pages offer a peek at our probable future as space merchants, miners, manufacturers, and settlers and how new space communities will emerge from a bleak frontier into a thriving archipelago of outposts, colonies, and settlements, each with a distinctive character defined by the people who live there, by the primary industrial sectors they support, and by the history they brought with them from their homes on Earth or from other places in the civilized solar system. Space will not magically emerge full grown but will build itself in fits and starts as new opportunities appear and are exploited.

While I was interviewing people with different ideas about space, I soon realized there is no single vision for space. Even common terms like "space settlement" take on dissimilar meanings for different advocates. I will try to parse these diverse visions into areas of commonality and difference so that there is some general understanding of the aspirations that different folk have for a settled space. Just as there are multiple ideas about what space settlement is meant to be and how it will come about, there are many assumptions about people living and working in space. As we will see, some of these basic assumptions come from generations of seeing space through the artificial theatrical lens of science fiction which often exaggerates and distorts basic science and human nature in convenient service to plot, character, and dramatic tension. Some of these accepted assumptions limit our view of space while others enable us to envision new possibilities that may be realized centuries into the future.

Of significant importance is the impact of space tech and commerce on Earth. Certainly, technologies developed for space will quickly migrate back to Earth and transform old ways of making and distributing goods. Commercial impact is inevitable. Beyond the impact of new space technology innovations on business practice, there is an even greater potential influence of how people living in space will change the way our species thinks about itself. We will carry our human evolution with us to space. The same evolutionary force of change and adoption of amoeba, ants, and our ape ancestors.

Still unanswered is if our historical tribal tendencies will continue with us as we migrate to space. Will we set aside conflict as a means of seeking power and acquiring wealth? Will we strive to abandon the imposition of scarcity and acquisition on the human condition? Will access to the near limitless availability of resources liberate us from the ceaseless flywheel of striving for more? Will the knowledge that we can traverse our solar system unleash a greater sense of shared purpose? Will we come to see ourselves as guardians of this new home or merely transients passing through on the way to another frontier and the promise of material wealth? Will differences of gender, race, and spiritual belief finally become inconsequential?

21 Questions About the Space Economy

Here is a list of some of the major research questions I explored while writing this book. Frankly, I did not expect to find definitive answers to many of them. My hope was to use these questions as a tool so that I could learn more about the topic. It turns out I discovered much more than I originally hoped. Before turning the page and diving into the rest of the book you may want to go over this list for yourself just to see how you would answer these questions. When you finish the book, you can circle back and compare your before/ after responses.

1. When will the new space economy officially begin? 2040? 2100? Now? Never?
2. What role will humans have in space?
3. Are prevailing notions of people living in space settlements realistic?
4. Given the hostile nature of space will robots or humans do most of the work?
5. Will there be a time when dominant economic activity will shift from Earth to space-based markets?
6. What will be the impact of the vast abundance of space resources on Earth's global economy?
7. Other than mining, manufacturing, tourism, energy generation, and ancillary supporting operations, what will be the dominant industries in the new space economy?

8. Will the economic development of space conform to a universal master plan or will space develop organically, driven by the priorities of countries, companies, and ambitious individuals in pursuit of self-interest?
9. Is there a predictable series of adoption phases that may be used as a guide for a company planning the best time to establish an enterprise in the new space economy?
10. What are the business models best suited to each adoption phase that will enable a company to better structure their business plan?
11. Will the greater space economy expand quickly across multiple locations in the solar system? Or will the industrialization of space follow a slower, controlled, more deliberate trajectory?
12. Will the development of the space economy redefine the relationships of sovereign nations with each other and with dominant space enterprises? Will future industrial and national space partnerships ignore the implications of traditional political borders on Earth?
13. Will supply chains develop independently (organically) or will they be choreographed by organizations wishing to control the direction of the new space economy?
14. Will space-based manufacturing technologies impact (disrupt) long-standing Earth-based business?
15. How will treaties, accords, and other well-meaning agreements be enforced in space? Who will enforce and adjudicate disputes in space when the space laws of one country conflict with those of another country or a private enterprise?
16. What is the prevalent/dominant reason for humans to go to space? What is the "right" reason? Are these two perspectives in conflict or in agreement?
17. Will the new space economy eventually become independent of Earth's economic support? If economic independence is reached, then when will this occur? Will Earth and space become economic competitors?
18. Will it be today's engineers or tomorrow's managers who will lead the way to a fully formed space economy?
19. What is the downside—the reason(s) to reject—the coming space economy?
20. Is there something to learn from sci-fi literature about how the space economy will develop?
21. How can individuals best plan for a career in the space industry?

What comes next is what I learned about the new space economy: answers to these questions and other surprising discoveries about the cosmos economy.

References

Andrews, E. (2019, June 26). *Who were the luddites?* History. Retrieved from https://www.history.com/news/who-were-the-luddites

Bell, J. S. (2015). *Super structure: The key to unleashing the power of story*. Compendium Press.

Krasavina, L. N. (2010). *Colonial trade*. TheFreeDictionary.com. Retrieved from https://encyclopedia2.thefreedictionary.com/Colonial+Trade

Rogers, E. M. (2003). *Diffusion of innovations*. Free Press. (Original work published 1962).

Vonnegut, K. (1980). *Player piano*. Delta Trade Paperbacks. (Original work published 1952).

Wikipedia Contributors. (2019, April 15). *Inductive reasoning*. Wikipedia; Wikimedia Foundation. Retrieved from https://en.wikipedia.org/wiki/Inductive_reasoning

Part I

The Business Case for Space

Beyond political, military, and scientific reasons to go to space, there is a growing list of financial incentives, and disincentives, for investors, sovereign states, and adventurous private companies who will seek profit and influence in the new cosmos economy. This section looks at the scope of the emerging space economy, the importance of the frontier in the process of industrializing the solar system, the role of human labor vs. robots in space, and the driving forces that will enable the economic growth of space.

2

The Cosmos Economy

The great thing about fulfilling your destiny in our modern, fluid, make-it-up-as-you-go-along world is that your destiny is something you can control. Heraclitus famously said that, "character is destiny," (ἦθος ἀνθρώπῳ δαίμων) by which he meant that a person's destiny is not a matter of predetermined fate but is something of her or his own creation (Wikimedia Projects 2004). In our modern world destiny isn't an inherited life ordained by birth, it is the result of initiative, ambition, courage, and drive. You are free to define your own success in life.

My best friend descended from a family of Southern slaveholders. It was something that gave him no pride, no honor. For him it was just a random fact of his history to reject and discard. When I think of Buzz I don't think of the stigma of his long-ago heritage, instead I think of his hard-won international reputation as an accomplished research scientist responsible for wiping out several auto-immune diseases during his too-brief lifetime. I don't think about the sins of his forbearers that might have limited his personal view of who he was or what he could accomplish, nor should you. He chose to step away from his heritage rooted in cruel self-interest and instead follow a different destiny of selfless service of his own making.

My old friend chose to devote his life to alleviating human suffering instead of allowing the stigma of his family history to define him. As a result, he was able to create a life that had a positive impact on thousands of others. For Buzz, destiny was a matter of individual choice.

What was true for Buzz in his life is also true for our greater society. We have a collective say in how we define success, how we act, who we are, and who we wish to become. The great thing about destiny today is that it isn't predetermined and immutable, we can sever our ties to the bad apples on our family tree, redefine our collective selves, cast aside old limiting notions of what is impossible, and embrace the life and destiny we want. This is how it is for mankind's new clear destiny in space; courageously casting off the old notions and embracing a new vision of our future.

Mention space and you will invoke a broad variety of images and ideas. Some people see space from a purely scientific perspective, a celestial spreadsheet full of formulae, physical laws, and cold facts. Others see space as a majestic firework display of great distant galaxies, super novae, black holes. Some see space through the lens of science fiction, zipping around faster than light, hobnobbing with alien species (may the Force be with them). There are those who think of space with childhood's wonder, a sense of marvel and curiosity about billions of stars seen for the first time on a clear summer night. Business folk see the future of the space economy through a lens of optimism and hope—and opportunity. Space is all those things; much more than any one of those things.

Today there are millions of people on our planet who believe space is a viable place to settle and live, a new home for mankind. Individuals, corporations, and nations may look to space with different agendas, but they share a desire to make space accessible, habitable, profitable. Some see space as a refuge from our suffering planet on life support, others see space as a new economic frontier with the promise of new wealth and a new path for mankind. There is both conflict and harmony in these differing goals, but the common thread is the anticipation that mankind is on the threshold of leaping off our home planet to settle the solar system with permanent communities, outposts, colonies, and industrial stations. Settlement advocates believe this off-world leap will help find new solutions for old problems, new beginnings for our adventurous species.

There are profound distances between the different extremes of space. Space is near and far away all at once. Space is the very long space-time it takes to travel in it. Space is the great empty gaps between planets, moons, asteroids. Space is the difference between what we know about it and the larger domain of what we have yet to learn. Space is distance from today's space activities to our future in space. There are spaces between safety and danger. Spaces between today's planning about space and tomorrow's actions to fulfill our dreams about space. It is all those spaces that make settling the cosmos the adventure it is for the new pioneers who will make it happen. This is about crossing the gap between thinking about space settlement to realizing the dream of mankind's future home in space.

The biggest space of all is the chasm between thinking we have no control of our destiny and creating a new destiny of our own design. A destiny that answers questions about who we, where we want to go, why we want to be there. Just ask the people who knew, admired, and loved my dear friend Buzz about living a life in service to a noble vision.

A Timely Topic

We are going to settle and work in space for the same reasons ambitious adventurers and forward focused nations have always gone to uncharted lands: the promise of profit and power, the lure of political expansion, the survival of the tribe, curiosity about the unknown, the acquisition of riches, a wish for a new life, and just because it's there.

Space is about new careers and ventures for professionals and entrepreneurs, it's about all the businesses we know and it's about all the new enterprises yet to come. Space business is more than GPS satellites and scientific experiments. The space-based economy is about finance, infrastructure investments, data analytics, habitats for space folk, mining and refining raw materials harvested in space, robotic manufacturing, industrial agriculture, and a host of ventures yet to be customized and adapted to the new high frontier. The new race to control and industrialize our solar system offers the prize of wealth from mining asteroids, building a new industrial economy, feeding millions of people with huge automated space farms, and establishing thousands of new outpost communities throughout our solar system.

Space today is a high-stakes dash to industrialize our solar system with the promise of financial and political power. The space sector has attracted billions of investment dollars from sovereign states, venture capital firms (VCs), private equity investors, new entrepreneurs, and established corporate players. In our super competitive world of winners and losers China is today's odds-on frontrunner to dominate in space, but don't count out the other countries, companies, and private investors who are also committed to commercial success in space. The potential for a huge profit from space is too good to pass up.

Space Angels, a leading venture capital firm focused on funding startup ventures in the space economy, reported a 73% increase in space-related investment in 2019. This represented a total of 269 ventures and an overall dollar value of $25+ billion. No matter how you value the over/under odds of the new space economy, by mid-century, when large-scale infrastructure projects begin to come online, the value of space-based enterprise is expected to exceed a collective market cap of over three trillion dollars. In their 2018 report The US Department of Commerce estimated the space economy will grow to at least 3 trillion dollars in just over two decades. A more conservative earlier projection by Bank of America/Merrill Lynch (2017) pegged the space market at 2.7 trillion dollars by 2045. After mid-century, when large-scale space infrastructure projects begin to come online, the value of space-based enterprise is expected to accelerate even more. These projections are no secret to the growing cadre of nation states, VC investors, and corporate planners (many not usually considered to be traditional "space" companies) that are adding space businesses to their long-range strategic planning projections.

Plotting the Course Ahead

Of keen interest is how entrepreneurs, companies, and countries will shape the developing space marketplace in the decades ahead. What follows is a look at the coming new space economy in our solar system. Where we are now, where we are likely to go in the near term and in the long run, how the space settlement economy will progress along a development curve from frontier phase to a fully formed trade economy, what the probable business sectors are that will prosper at each phase of development, and the impact of robust space settlements and thriving space economies will have back here on Earth. More important, we will explore the prospects for individuals, entrepreneurs, investors, and private firms seeking new opportunities in space.

Here are a few themes and topics we will discuss to help address these critical questions.

(a) A look at how the developing space-based economy will permanently alter the social, political, environmental, legal, and economic landscape of Earth.

- Will new space industrialists compete with Earth-based industries? Will they control and exploit the solar system's mineral riches? Will space industrial output and innovation eclipse Earth's longstanding leadership role in the new trans-space marketplace?

(b) The value proposition for space. Why should investors, sovereign states, and ambitious private companies care about space at this early stage of development? Beyond the political, military, and scientific reasons to go to space, there is a growing list of financial incentives, as well as barriers, for building the industrial space economy.

- The days of US/NASA dominance in space are rapidly ending. Foreign actors like China, India, Israel, United Arab Emirates, Russia, and even Brussels have their sights set on staking a claim in the new trans-space marketplace. Who will be the winners of this new space race and who will likely carve the bigger slice of space for their own political, military, and economic agenda?

(c) A look at how past commercial ventures took calculated risks and built empires in the New World, India, and other undeveloped regions of the world. What are the lessons learned from those frontier ventures that may serve as lessons for new space investors?

- The new space frontier will have a similar impact as did the California Gold Rush and the Oklahoma Land Rush of the nineteenth century; new space investors and adventurers will seek early-mover advantage from exploiting space opportunities. After the major players carve out their share, what will be left for the late adopters and everyone else?

(d) A discussion of the economic risks and opportunities in the new space economy, including the role of humans vs. robots, infrastructure challenges unique to space, and which industries are most suited to off-world development.

- Man's biggest enemy in space is space. It is a hostile place; it is no country for Earthmen. There is no infrastructure in space, no platform upon which to build the new space economy, no network of suppliers, and no consumer base (yet). How will early space industrialists create and follow a shared strategic roadmap based upon mutual assistance?

(e) The ripple effect. How will the new space frontier impact traditional market activities back on planet Earth? The new financial space race for economic dominance of the solar system will pit competitive nations and companies against one another, new alliances will form, old trade agreements will be renegotiated, partners will become competitors, and former friends may become adversaries.

- How will the threat of the new space economy's potential to surpass longstanding Earth-bound production technologies and redefine the scope and influence of Earth's economy?

(f) How can entrepreneurs, investors, and individuals participate in the emerging industrialized space economy?

- What are the risks and rewards for new entrants adopting to new phases of economic development? What are some best-fit business models that will likely thrive in each phase of adoption? How can an entrepreneur leverage the predictable phases of development to their advantage?

(g) The historical development of the Western Hemisphere and third world economies are often referenced as a template for the development of the new space economy. But space is not the American West of the 1880s and rockets aren't the same as Conestoga wagons or railroads, yet there are many mini lessons from the past that are worth visiting.

- What are the key lessons learned from the past that we can apply to the development of the new space economy?

Why Industrial Space Won't Be Business as Usual

This is a look ahead to the emerging business of space. It is where space, business, and technology join to create a new vision and answer critical questions for professionals, investors and passionate observers who are curious about the future convergence of space and business. The U.S. Commerce Department currently estimates that over 70 countries and 1500 private corporations are presently active in the rapidly growing space economy. This global commercial interest in space business is just the beginning. My research led me to some startling and often controversial conclusions about what the coming space economy will look like and how it will dramatically impact the future of our planet. A preview of some of these findings that will be discussed include:

– Only very few select nations, investors, and companies will control the new space frontier. They will likely come from the ranks of early adopters and aggressive investors willing to take the risks necessary to capture substantial rewards.
– Most of Earth's population will not likely benefit from commercial activities in space in the short term. Early adopters will reap tremendous advantage due to the principle of cumulative advantage which means the *haves* will have more, and the *have-nots* probably won't.
– Space is not a place for people. But if you are a robot, cobot, or an AI-driven industrial machine, then there's no place like space.

- Space civilization will not be civilized for some time. Space laws will be virtually unenforceable in lawless space in the early chaotic stages of economic and social development. Space treaties and policies will likely be ignored or "reinterpreted" in the inevitable struggle to control space.
- High barriers to entry mean only companies, consortia, or countries with deep pockets will play and prosper from their first-mover advantage.
- First-movers who focus on infrastructure will have a long-term sustainable competitive advantage.
- The development of industrial space will not solve all of Earth's Ills. Our social, political, economic, environmental, and military problems will continue with or without the new high frontier of space.
- The down-mass barrier. Unless rocket makers develop the capability to move huge payloads of finished goods from space factories down to Earth's markets space-based manufacturing may stall. We are a consumer-driven economy and Earth is where all the traditional consumers are—for now. The silver lining of this problem is the rapid development of a robust B2B space-to-space market.
- Significant threats to establishing new space markets are (1) corporate bureaucracies that tend to stifle entrepreneurial creativity and risk, (2) a near-fanatical devotion to short-term (i.e., quarterly) results, and (3) entrenched industries on Earth who see new competitors from space as a threat.
- There is no unified master plan to develop the space economy. The only "roadmap" is a jumble of competing interests and policies designed to leverage influence and power for the advocate nation or organization. This glaring gap creates an opportunity for strong political leadership.
- Finally, the high stakes of industrial development in space will have a permanent influence on Earth's economies, political systems, and citizenry. The new cosmos economy will challenge our eons-old Earth-centric chauvinistic thinking about centralized power and will unleash a tsunami of growth throughout our solar system for centuries to come.

It's About the Journey—Not the Destination

Think of an immense unexplored region filled with new worlds not yet discovered, still unmapped. An enormous place made up of a nearly endless collection of welcoming and hostile places, a hodgepodge of large worlds and mini moons. As surveyed by NASA (2019) this territory is rich with many millions of tiny space rocks, millions of asteroids, hundreds of minor planets (like Pluto), over 350 moons (175 of which are wedded to a major planet), and of course the eight major big-boy planets we think of when we think of our solar system. Throw in the asteroid belt between Mars and Jupiter and then add the Kuiper belt beyond Neptune where Pluto lives. This is the virgin terrain of our solar system where space settlement and commerce will happen, a tabula rasa of civilization's next big step.

It's not what you don't know, it's what you don't know that you don't know. At this moment in time we generally acknowledge that there is much we don't know about our space backyard. But there is much more we don't know about what we don't know. This lack of assurance about space can easily undermine confidence in the settlement enterprise. The best way to solve the problem of not knowing about what's out there in our solar system is easy: Go there and find out!

Ancient seafaring myths and superstitions held that an incautious captain might steer his ship off the edge of the Earth and would be forever lost in the abyss. Our solar system is so large and virtually limitless that tomorrow's inattentive space captain may find a similar fate in the far reaches of space. But, as suggested by Markham's article about the process of discovery, *Do You Know What You Don't Know?* (2012), instead of falling into the abyss she may discover something new just beyond the edge; something she didn't know that she didn't know!

References

Contributors to Wikimedia Projects. (2004, March). *Heraclitus pre-socratic Greek philosopher*. Wikiquote.org; Wikimedia Foundation, Inc. Retrieved from https://en.wikiquote.org/wiki/Heraclitus

Markman, A. (2012, May 3). Do you know what you don't know? *Harvard Business Review*. Retrieved from https://hbr.org/2012/05/discover-what-you-need-to-know

Tran, F., Nahal, S., & Ma, B. (2017). *To infinity and beyond—global space primer*. Bank of America Merrill Lynch (Thematic Research).

3

Blue Oceans and Greenfields in Space

Investment in the cosmos economy is already experiencing growth in a wide variety of areas including energy production, communications, data analytics, space-based manufacturing, mining, finance, construction, and tourism. Private companies, not just nation states, are playing an increasing role in building the new space-based economy. Market timers who study the potential returns from investing in space recognize the best time to invest is now, before the rest of the mob jumps in. Kim and Mauborgne in their book, *Blue Ocean Strategy*, (2005) noted there are successful firms in business today that didn't even exist just 20 or 30 years earlier. Think of Starbucks, Tesla, SpaceX, Amazon and products like Apple's smart phone, Microsoft PCs, cloud services, and Google's search capabilities. These firms saw a new market where none previously existed and created a blue ocean for themselves. Blue oceans all, they were market timers who glimpsed the future and acted ahead of the crowd. Such is the case for business investment in space.

Laying the Foundation for Space

It may also be appropriate to think of the space frontier as a greenfield market ("greenfield" is loosely defined as an untrammeled, unexploited, untapped commercial opportunity ripe for development). It's not just companies like Boeing, Mitsubishi, OffWorld, Bigelow, Planetary Resources, SpaceX, and Amazon who are serious about space commerce. In the new race for space primacy nations like China, India, Russia, Japan, UAE, Belgium (yes, the kingdom of Belgium) and the USA also plan to make their mark and define the new space economy in their own terms for their own national self-interest. Corporate planners, entrepreneurs, and visionary venture capital investors (VCs) are weighing the costs and benefits of when to enter the market, which business model has the greatest probability of success, and how to build a long-term sustainable competitive advantage in space. The greenfield market in their sights is the high frontier of space beyond low Earth orbit (LEO) where current commercial space activities are taking place. The untapped solar system is the blank page waiting for industrialists and investors to write the next chapter of Earth's economic growth.

Success for the space investor will be measured in long-term profitability and economic sustainability. As the volume of space settlements and outposts continues to grow there will be more opportunities for commerce, more wealth generated for investors. The traditional metrics of return on investment (ROI), corporate earnings, brand dominance and reputation will continue to be the pulse taken of a firm's financial success, but other indicators will also speak to the well-being and fitness of the space enterprise such as competitive strength, rate of expansion to new space markets, new products, business line extensions, and agile adoption of business operations to align with the changing commercial space environment.

Not to be overlooked is the agility of the leadership team. In a fast-paced dynamic market, it may be convenient to assume early success will sustain for the long haul. But this is not always the case. Early achievement may create a false sense of complacency about the long-term viability of the business model, the strength of the market being served, and even the future potential the business given that the space playing field will likely change unpredictably. In this setting it is critical for leadership of the enterprise to willingly pivot to new market opportunities and abandon yesterday's formula for success as new opportunities arise. This is a highly volatile business setting. Those unable or simply unwilling to ride this erratic roller coaster will probably not last.

Other measures of a firm's success are more subjective, more subtle. Many founders of companies have a heartfelt sense of social responsibility, political activism, environmental sustainability, or simply legal access and equity concerns. Professor Edgar Schein famously observed in *Organizational culture and leadership* (2010) that the ethical values of the founders often determine the culture of the company and the types of people who want to work where those values are important in the long run. Beyond the traditional bottom line of revenue minus expenses is a more complex and ethically oriented accounting known as the triple bottom line.

To achieve a profitable bottom line in this model, the firm must accept a decidedly un-businesslike approach to profitability by incurring additional costs beyond the costs of production. The triple bottom line adds the costs of social responsibility and the costs incurred in support of environmental sustainability. These added costs aren't hypothetical; they are real costs. This approach comes with an organizational commitment—supported by company leadership—to invest in socially responsible activities that serve the greater good and to make investments in the preservation and perpetuation of a sustainable environment. Not all companies can keep this promise. Not all investors see this business model as a viable path to getting a rapid return on their investment.

An example of a company who keeps a social commitment is Zappos Shoes whose business model is to donate one pair of shoes to a disadvantaged community for every pair sold. Obviously, this increases costs and decreases profit margin drastically. For those who look to the space frontier as a platform for a new ethical beginning of commerce the challenges of creating and sustaining a triple bottom line commitment will be significant.

There is little doubt that the development of space will have a broader impact beyond seeding economic growth and human settlement. The social, political, and legal environments on Earth and in space will be altered by the weighted influence of new frontier activities. The role of citizens in space will be defined by their efforts to carve a new life, a new society, and a new definition of personal achievement and success.

How we Imagine Space

Attitudes and ideas about space—what it is and what it's potential role in our civilization—have evolved from mysticism to fictional speculation and, more recently, to an achievable dream in the minds of people all over the world. Dr. Alex MacDonald noted in his book, *The Long Space Age*, (2017) that the history of humans and space has been in the making for millennia. It is only in the last half century that we've come to think of space as something real, an attainable destination.

At the core of my research is the nagging notion that any study about the future is fiction. And if the inquiry is about the future of space, then it must be to some degree space fiction, AKA science fiction. But that well-established genre comes packed with baggage I am unable and unwilling to carry. I know where the story starts but can only guess about where it will go. I hope it has a happy ending for the hero and an unhappy ending for her enemies but there is no telling now. All I can be sure of is that this story will fill volumes, have lots of surprises along the way, and will come to set the stage for all sorts of stories to come.

Much of our early notions about space came from science fiction. Jules Verne (*From the Earth to the Moon*), H.G. Wells (*War of the Worlds*), Isaac Asimov (*I Robot*), Ray Bradbury (*The Martian Chronicles*), and others wrote stories that broke away from formalist fiction and presented readers with the prospect that the future was unknowable, exciting, and full of possibilities. Science fiction seldom fully explores the future dimensions of space commerce (although it often lurks unspoken and unseen in the background). The cold gray facts of commerce would have probably muddied the narrative and made the hero a bit boring. Later authors like Robert Heinlein (*The Moon is a Harsh Mistress*) and Frederik Pohl (*The Space Merchants*) came closer to acknowledging that people in space needed to be busy doing something other than just having adventures and pursuing their quest for justice. Their books acknowledged a more complex story setting that often-revealed political intrigue and social disparity.

Professor Lisa Yaszek of Georgia Tech, a leading scholar on science fiction's impact on society, maintains that most well-constructed science fiction is based upon a foundation of an economic or commercial backstory that drives the plot and enables character development. As examples she cites *The Man Who Corrupted Earth* (1980) and *The Privateers* (1985) where, "entrepreneurs take charge of the conquest of space after government has given up." (It seems like current trends to privatize space and reduce government control are just catching up with earlier science fiction literature.)

Science fiction movies of the 1950s (e.g., *Forbidden Planet, The Day the Earth Stood Still, This Island Earth*) and mid-century TV shows (e.g., *Star Trek, Twilight Zone*) made efforts to inject a universal humanism into the plots. Later popular films (e.g., *E.T., Star Wars, Close Encounters of the Third Kind, 2001: a Space Odyssey*) brought science fiction into the realm of mainstream media. All these fictional characterizations of life in space anchor our expectations for what actual life in space will be. When people think of space, they think of a place made safe by flawless technology because sci-fi tech is flawless, where the life-threatening problems of deadly radiation, the effect of weightlessness on the human body, and the threat of imminent death from lack of air, water, or food is of little concern because all of these problems have been resolved (or ignored if their discussion doesn't serve the plot). In this way our image of life in space is influenced by the ideal of fictional fantasy.

But fictional space is not real space.

Some dangers of operating in space are well known; sudden bursts of cosmic rays, solar flares, the debilitating effect that living in zero gravity has on human tissue and skeletal strength, not to mention the long-term impact on cognition, and the unknown influence of living in space on human reproduction. None of this is good news. Add the probable prospect that there are alien beings—microbes? Monsters?—and the utopian hope of homesteading your little house on the asteroid becomes a rock too far. There are also potential health benefits in space; the aged and infirmed may benefit from a life with less gravity and thus less stress on their joints, muscles, bones, and grandkids.

Yet we are willing to suspend our knowledge about the dangers of space in favor of preserving the rosy vision of happy and confident space folk stridently forging a new world out in the cosmos; a sort of bold space utopia where technology takes care of everything, where magical machines make food, where there is no need for conflict because scarcity has been abolished and everything is provided. It's a better vision than the reality of living in space and for many sci-fi fans their space fantasy is a better vision than the harsh reality of living down here on Earth. But it won't happen in our lifetime and probably won't happen at all. Space will be a hard scrabble place of ambitious, aggressive folk trying to create a better life by leveraging whatever talents and resources they have in a harsh setting that isn't made for them and doesn't want them around.

Then why tempt fate? Why bet treasure, reputation, and life on the long shot of making a living, making a life, in the hostile space frontier? For some with nothing left to lose the answer is "why the heck not?" If I lose, I'm no worse off than my prospects here on Earth. Such was the "make-it-or-break-it" mindset of countless migrants who have flocked to America for centuries. For many it was a gamble that paid off in the long run, for their descendants in the next generation of two. That is the bedrock attraction of being a pioneer in space; it's not working for me here so I might as well take a gamble on someplace else, even if that place is across the globe or out in the space frontier. It isn't just the idea of *going* to a better opportunity, it's also *abandoning* an old life that isn't working the way you expected, a life bereft of the opportunities you desire that will make your personal vision of success come about. In time, space may be the new chance at a better life for everyday folk who see a life in a cosmos settlement akin to a new start and a chance to be part of something important.

But wait, isn't the news full of stories about billionaires investing truckloads of dollars in new space ventures? What about the top one-percenters who are funding space ventures? They're motivated by something more than just finding a path out of persecution and poverty to seek a better life. They're motivated by the prospect of big paydays, to stake a claim in the high-risk space sector. But for most of today's space entrepreneurs their enterprise is much more than just profit. It is a chance for those who have accumulated wealth to put their good fortune to work in pursuit of making something better for humanity; An act of altruism, a chance to make their dreams of space come true.

Crashing the Billionaires' Club

When we think of the people most associated with space we rarely think of the nameless engineers, financiers, and middle managers who do most of the heavy lifting. We think of celebrity names like Elon Musk, Jeff Bezos, Richard Branson, and a close cadre of marquee business personalities who are constantly in the news. There are lots of other investors and space entrepreneurs, of course, but members of the billionaire club continually grab the spotlight. This may be partly due to our celebrity-centric culture where being rich and famous is a measure of value. This may also be the result of savvy self-promotion: Elon Musk continually grabs media attention with a new business venture or a just-over-the-line comment. And when he makes the news, he sells more Teslas, builds awareness about his solar panel business, his lithium-Ion batteries, his underground transportation tunnels, his reusable rockets, and his space-based internet for the planet. Being in the news is good business. Likewise, Richard Branson is not above posing with swimsuit models and taking a trip to space as the first passenger in one of his Virgin Galactic space planes. The space-related efforts of Jeff Bezos' Blue Origin are less about public grandstanding and more about creating space-related transport technology.

Each, however, has a strategic vision of where they want to take their ventures. Richard Branson makes no secret of wanting to further develop his space plane technology to compete with current long-haul airline routes: A 15 h flight from the USA to Japan could be done in just a few short hours. Elon Musk makes clear that his goal is Mars and has already picked the spot where he wants to be buried on that planet. The foundational efforts made today, and the large sums of investments that have migrated to these enterprises, speak to a longer-term view about the development of space settlements. The real motherload in space may not be from gold or other rare minerals mined from asteroids, but from generous expected investment returns made from new space economy enterprises in operation today. Taking the enormous risk of being an early mover has the upside potential of generating high returns. Who better to take advantage of this opportunity than wealthy nations, companies, and ambitious citizen investors who are willing to play the long game by getting in on the ground floor?

What's Holding Back the Space Economy? (Hint: It's You)

Will humans settle space? What role will people play in the space frontier? Will the dream of living in space be forever just out of reach for space enthusiasts no matter how willing they may be to go there and live? Space will not be kind. It will disappoint. It may even betray your best intentions. Space, like a pandemic, will make life hard and strenuous. Space will deprive men and women of all the convenient Earthly pleasures of breathable air, comfortable climate, shelter, freedom of mobility, abundant access to food, and social interaction. Life in an early space settlement will be like a life of captivity in a Siberian Gulag prison camp where the inmates are dependent upon each other and the limited external supplies for most of their life support, where the notion of freedom and independence is whittled down to a narrow scope of what the prisoner can do without rocking the boat and risking injury or death to themselves or to others in their closed community. Living a survivalist life in space may not match the utopian vision that tomorrow's space pilgrims may have. Social skills and the need to balance personal imperatives with those of the rest of the cohort will be critical. The importance of developing a sophisticated skill set of social behaviors, known as emotional intelligence (EQ), may be the key to living in close quarters in temporary space transit and in a permanent space settlement. The ability to get along with others may be the most critical talent a settler may possess. Daniel Goleman's book *Emotional Intelligence* (2006) presents the idea that human interaction skills can be taught and are the key to success in groups. Life on a far-flung space settlement may be best if taken in short stretches of only a few years instead of a lifelong sentence.

Before people arrive on the scene robots will do the heavy lifting to set the stage and establish basic infrastructure. The Chinese employed this multiphase process when they established their lunar research station in 2019 with *Chang'e 4*. The use of robots to build structures and infrastructure raises the question of why people are needed in space at all. If the purpose of establishing a space outpost is for purely industrial reasons, then most of the labor-related activities can be delegated to robots or some other automated process. The role of human managers would be to monitor robotic activities, make corrections, and initiate periodic maintenance via teleoperation. If the outpost is on the Moon this can be coordinated from Earth or from a cislunar installation. The further out in the solar system, the greater the need for human overseers to be near the action to allow for longer communications lag.

This means that the first permanent space settlers will be robots. They can work 168 hours per week without vacations or bathroom breaks. If they break down, they can be scavenged for parts. Robots have no need for housing or food or health insurance benefits. They can act as generalists or specialists depending upon their programming that day. The best part is they don't have all those messy human emotions, personality quirks, and ambitions that make running an organization so challenging.

References

Goleman, D. (2006). *Emotional intelligence*. New York: Bantam Books.
Kim, W. C., & Mauborgne, R. (2005). *Blue ocean strategy*. Boston: Harvard Business School Press.
Macdonald, A. C. (2017). *The long space age: The economic origins of space exploration from colonial America to the cold war*. London: Yale University Press.
Schein, E. H. (2010). *Organizational culture and leadership* (Vol. 2). New York: John Wiley & Sons.

4

The Importance of Frontiers

The Turner Thesis 2.0 …

In the summer of 1893, at the Chicago World's Columbian Exposition, Frederick Jackson Turner a minor academic in the new discipline of social science, presented his now-famous paper on the closing of the American frontier. What came to be known as the *Turner Thesis* (1893) forever changed thinking about the influence and importance of the frontier and how dramatic changes of the frontier profoundly impact future economic and social issues. His thesis was deemed so significant that it is still taught at colleges and universities today.

At that earlier time a World's Fair exposition was more than thrill rides, Ferris Wheels, side shows, and popular amusements. An exposition was a cultural benchmark event that made a statement about the state of culture and society. It was the custom then to also present scientific talks on a variety of topics from agriculture to manufacturing innovations. Some presentations looked to past achievements while others spoke about how future engineering wonders would make life better in the coming twentieth century. The common theme was how the future would offer prosperity and opportunities for common folk to live better lives in the new America being born on the prairie frontier. These talks were meant to be positive messages of uplifting optimism, hope and confidence about the future.

All roads led to Chicago. It was the commercial hub of the Midwest, a manufacturing powerhouse, and a place where influential tycoons and industrial robber barons established their headquarters. Significantly, Chicago was the central rail connection for East/West transportation, a destination for tens of thousands of Eastern European immigrants, and the place from which manufacturers shipped goods all over the world. The West Side stockyards turned the cattle shipped by rail from Texas and Kansas into new western wealth and transformed agriculture from small family farm producers into large-scale commercial businesses. By 1893 at the World's Columbian Exposition, Chicago was able to boast that it was much more than merely, "Hog butcher for the world, tool maker, stacker of wheat, player with railroads and the nation's freight handler; stormy, husky, brawling, city of the big shoulders," as Carl Sandburg (2013) later wrote in his poem, *Chicago* (1914). It was where the western prairie started and where the frontier finally ended.

The place and timing couldn't have been more significant. Chicago at the tail end of the nineteenth century represented the industrial might of American Manifest Destiny. Just three decades after the civil war had seen dramatic economic growth throughout the nation, especially the frontiers of the west. Immigrants flocked to reap the economic benefits and to start new lives in the new world where they could be free of centuries of poverty and persecution in the old world. There was no reason to believe there would be anything to dampen the dreams of an ambitious person seeking to carve out a prosperous life on the never-ending frontier. Everyone knew that the driving force behind the nation's economic prosperity was the western frontier. That was where vast reserves of natural resources were. That was where adventurers went to stake claims, start new businesses, make their fortunes. That was where the "empty" spaces gave promise to the notion that everyone with gumption and a dream could rise above their social station, no matter how low, and become part of the new settled middle class. All that was needed was the courage to travel west and create a new destiny. After all, it was common knowledge that the bulk of America's phenomenal economic growth had come from developing the vast untapped continent by turning virgin land into farms, prairies into cattle ranges, forests into timber fortunes, trading outposts and settlements into villages, and villages into cities with shops and schools. The mindset of investors near the end of the nineteenth century was that there was no end to the opportunities of untapped resources, that cheap land and open territories would be available forever.

But Turner's *Frontier Thesis* said otherwise. He told his shocked audience at his World's Fair presentation (the current site of the University of Chicago's Midway) that the American frontier was finished; the march of American civilization had come to a halt.

His message was that the prevailing image of the frontier as a vast open untamed territory where a wanderer could travel freely was now history. In fact, Turner reported, it was no longer possible to traverse the continent from

North to South or from East to West without encountering numerous rail lines, roads, and barbed wire fences that turned the open untamed prairie into private property. The useful land had been claimed, settled, carved up, cultivated, and civilized. The frontier was dead because the unlimited free range was no more. The wide-open spaces of the Old West were wide open no more.

Furthermore, maintained Turner, the frontier's closing was more than just a matter of commerce and economics; there were dire social implications. Turner observed that one of the benefits of an untamed frontier wilderness was that social misfits, malcontents, and outcasts (what we like to call rugged individualists today) could safely retreat to places beyond civilization and live free of social conformity because the frontier acted as a safety valve for a well-ordered society. As the frontier diminished and cities grew, greater numbers of people lived and worked in proximity with one another. The pressures of urban life, especially for those who could not or would not easily adjust to life with others in the city, often resulted in conflict and violence. Turner made other predictions stemming from his observations about the consequences of the end of the frontier. He noted that because there was no more untamed land there would be no more refuge for untamable people. With no place to go these society outliers would have to live on the urban fringes, without income or shelter, as wandering homeless pariahs.

The frontier had always been a haven for society's outsiders who eschewed living a civilized life in a city among others, like the mountain men and trappers who savored lives alone in the "uncivilized" wilderness. There were those who famously preferred the hinterland to civilization, like Daniel Boone whose myth said that he kept pulling up stakes and moving his family further and further into the wilds whenever he thought another settler might be within miles of his cabin. The closing of the American frontier signaled the end of true freedom for those who were seeking to carve out a new life, who had struggled and failed to fit in with civilization and adjust to the pressures of a formalized society. The closing of the frontier was more than a geographical phenomenon—it was a social issue, a political issue.

Turner took the implications of the end of the frontier even further. He maintained that it wasn't just the closing of the American western frontier, it was the end of western civilization's frontier. In *The Frontier in American History*, (1920, 2013) Turner later wrote about his observations on the impact of the end of the expansion of western civilization for America and the rest of the world. Turner boldly predicted that political and military confrontations would result from the lack of new opportunities for land acquisition and that world powers seeking to expand their influence would resort to large-scale conflicts to acquire vital new resources and territories from competitor states.

Without society's safety valve, Turner predicted, the end of the frontier would invoke decades of urban violence, social stress, increased criminality, and even world wars. The unrest and violence of the twentieth century unfortunately validated his predictions.

The process of carving out a civilized community from the untamed frontier isn't something that happened all at once. Turner thought of the frontier as, "the outer edge of the wave the meeting point between unmanaged chaos and tamed civilization." As observed in Peck's *A New Guide For Emigrants to the West* (1837) the raw wilderness becomes civilized in successive waves of settlers, each with their own motivations and capabilities. Writing extensively about the western frontier Peck observed that there were three major epochs in a settlement's maturity. Peck saw the development of frontier settlements, "… like the waves of the ocean, … rolled one after the other." The first wave was the pioneer who is a transient occupant of the range. "The pioneer is an opportunist seeking subsistence from the environment for his family. His purpose is autonomy, to be his own lord and to live apart from the control of civilized society. In time, when others start to settle in his realm, he moves on to another uncivilized range to repeat his process."

Peck's second wave of emigrant comes with more permanence in mind. He wrote that the settler's goal is to establish security for their family and to create a community with order and social identity. It is during this wave that local institutions like schools, small factories, stores, and municipal government are established. The third wave is one of speculation and economic opportunism. In this wave the new occupants come for a chance for wealth from industry and the acquisition of property, not because they want to create a close-knit community. They are advocates of growth for the sake of growth. They see the settlement in terms of the future, not in terms of its heritage. There are those who are not replaced by a successive wave. They will be the keepers of old values and origin stories—not necessarily the centers of power and influence.

Peck concludes, "There is not tabula rasa. [E]ach frontier did indeed furnish a new field of opportunity, a gate of escape from the bondage of the past; and freshness, and confidence, and scorn of older society. Impatience of its restraints and its ideas, and indifference to its lessons, have accompanied the frontier." His views represented the common romantic notion of the liberating freedom of the frontier; a place where the old ways could be cast off to make way for the new ways of thinking and being in the world. F.J. Turner referred to Peck's different phases of adoption when writing about the importance of the frontier to civilization and the dire impact of the end of the western frontier on greater society.

Often overlooked is Turner's concern about the influence of large corporate operations in the frontier. Because he was influenced by Peck's writing, he saw the potential of the frontier as a catch basin for the socially disenfranchised, an opportunity to escape their "have-not" social station and gain a foothold on the social ladder of success. For Turner the railroads, mining operations, grain dealers, and other large industrial companies threatened the economic access of small independent farmers and budding mercantile operations to the open opportunities on the free range of the Western frontier. Frederick Jackson Turner may have accurately predicted the impact of the end of the American frontier and extrapolated the consequences of this singular American event to the end of an epoch in human civilization, but in space we have discovered a new frontier.

Turner was right for his time, but he couldn't have foreseen how the coming of the new space economy would open new frontiers across the solar system.

The New High Frontier

The boundless expanse of undeveloped space will enable new frontiers in our solar system to germinate, flower, and replace Turner's cautionary assumptions with renewed optimism about our collective future. The new frontiers of space will be a story of water, minerals, far-away outposts, new city-stations positioned in cis-planetary deep space and of human settlement. Some frontiers will be highly civilized, others permanently primitive and crude; a full spectrum of habitation all happening at the same time. A bouquet of possibilities.

The U.S. western frontier blossomed because ambitious, tough-minded visionaries with dollar signs in their eyes saw opportunities for riches and took great risks to make their vision come true. Some famous frontier names, like Levi, Stanford, Getty, Huntington, and others have survived and become venerated models for beneficence and success. (Others, like Billy the Kid, Butch Cassidy, Doc Holiday, and their ilk became notorious for less-than-honorable reasons.) The new space economy will follow a similar path of bold (or crazy) entrepreneurs, visionaries and risk-takers (like Bezos, Ballmer, Branson, Bigelow, Brin, Musk, and others) who will transform the frontier of the new economy into the dominant marketplace from the twenty-first century onward. Maybe a century from now we just might remember some of their names, too.

Why Is this Space Economy Different from all Other Economies?

When I asked experts what they thought the new space economy would be like and what living and working in space would be like, I got long pauses, puzzled looks, and blank stares. Was I asking something they never thought about before? Occasionally someone would respond, "Why would anything be different about doing business in space?" Why indeed. When Columbus pitched his "new-route-to-India" investment scheme to the royalty of Europe (and got rejection after rejection) I imagine he had trouble explaining his vision of exploration and riches when he was asked, "Why would sailing your ships to the west and risk falling off the edge of the world be a good investment?" Why indeed.

A convenient way to examine the potential impact of the new space frontier is to compare it to the old American frontier. Think: cattle barons, Dodge City saloons, 1849 gold prospectors in the Sacramento Valley, Sears's mail-order catalogs in the 1890s, and homesteaders desperately hoping to carve out a life in the middle of nowhere. All those hopeful industrialists, adventurers, and social misfits came to the west to create their individual vision of something new. While you're at it, also recall violent range wars, summary mob justice, claim jumpers, robber barons, the decimation of native peoples, and how the myth of Manifest Destiny gave license to unbridled economic expansion at the expense of the environment and those without influence or power. Both good and bad results may come from the forge of creating something from nothing.

Turner saw the frontier as more than a physical location. He saw a boundary between the past and the future, where people could escape from the pressing imposition of an over-regulated social structure. He acknowledged that unsettled frontiers are more than places of refuge. They are also beginnings of new norms, new social constructs, new values, new collective stories that shape a culture, and new opportunities to define success in your own terms. *The Turner Thesis* (1893) is important for the imminent space economy because it acknowledges the importance of the undeveloped wilderness in civilized society as an incubator for hatching civilization's next incarnations. Turner's observations about closing Western civilization's last frontier are prescient because the space economy will rekindle the old frontier ethos and create new opportunities for social change, economic innovation, growth. And hope.

4 The Importance of Frontiers

The cosmos economy isn't going to happen all at once. It will not hatch from its incubator fully formed and ready to gallop. On the way to creating a wholly mature market space commercialization will progress through at least five discrete phases of adoption. The first of these phases is a phase of development characterized by innovators and risk-takers: the frontier phase.

There are some assumptions that are critical for the entrepreneur who is planning to launch a new space venture or a venture capital analyst who is evaluating the profit potential of a new space business investment. Either way it's important to fully assess the upside and downside of the proposition to get to know the lay of the land. Table 4.1 is a high-level comparison of basic assumptions about the old American frontier with assumptions about the new space frontier economy. It is a convenient way to set the stage by looking at the striking similarities and differences between these two extreme frontier settings.

Brink Lindsey, writing on *Frontier Economics* in *The American Interest* (2011), observed that, "the key distinction [of economic expansion] is between growth as more of the same and growth as something new and different." The danger of relying upon more of the same is that both markets and production capacity soon reach their peak where demand diminishes and production economies plateau. Lindsey notes that growth by doing more of the same, *imitative* growth, "occurs within the existing technological frontier; the latter [growth as something new and different], or *innovative* growth, pushes that

Table 4.1 This summary assessment of old and new assumptions about a frontier economy shows that while some elements persist (e.g., cost of transport, exploitation of natural resources) the impact of technology and big corporations will have a greater effect launching the space frontier economy

Comparison of primary business assumptions about doing business in a frontier economy

Assumptions old and new

Old US Frontier Market	**New** Space Frontier Market
1. Plenty of cheap labor	1. No cheap [*human*] labor
2. Little need for high tech	2. Tech is key ingredient
3. Small family enterprises	3. Large corporate investors
4. Agriculture based economy	4. Tech, mining & mfg. to start
5. Low barrier to entry	5. High barrier to entry
6. Unchecked exploitation of natural resources	6. Unchecked exploitation of natural resources
7. Centralized economic focus	7. Decentralized econ. focus
8. Econ. Dev. likely influenced by local actor politics	8. Econ. Dev. likely influenced by sovereign actor politics
9. Individual labor autonomy defines economy	9. A.I. & robotic autonomy defines economy

10. Is space the end of Dr. Frederick J. Turner's *"Frontier Thesis"* (1893)?

frontier outward." That's the essential promise of the emerging space economy. Thus, it is likely the space economy will first grow by innovation before it grows by doing more of the same via increasing scale.

Years ago, I had a grad student who told me how she coped with the pressures of her life of no sleep, being on call 24 h a day to her dissertation chairperson, and basically having no life of her own. To escape her situation, she drove her car aimlessly around L.A. for hours until she got lost (this was before every car had GPS or easy access to Wayz). Getting lost was her liberation. Getting lost untethered her from her pressures of school, work, family. Getting lost sharply shifted her attention away from her day-to-day life and forced her to refocus on the process of getting *un*lost.

Chances are that space will disconnect explorers and adventurers from their reality very quickly. They won't have to go far to find themselves upside down and inside out or up against their fundamental core assumptions about what is and isn't important. Space will gladly serve up sudden and scary liberating moments from reality and a drastic escape from whatever it is that people in space hold holy and dear. My grad student would love space. She had an adventurous spirit and a fascination with going to the edge and then finding her way back without breadcrumbs to guide her. Space people take note.

Healthy economies rely upon growth to keep them vibrant and competitive. In the past, economic expansion came via epochal changes like the development of organized agriculture, the Industrial Revolution and more recently, the impact of artificial intelligence (AI) and robotics. These monumental socio-economic shifts made plantation labor obsolete, introduced mass production, connected global commerce via the internet, and created whole new industrial categories based upon technology and information. The next opportunity for growth, the coming space-based economic revolution, will change the nature of business and transform our planet's economy. Space investment and entrepreneurial opportunities are just over the horizon.

Figure 4.1 shows the four fundamental areas of space activity. They are (1) Scientific exploration, research, and experimentation, (2) Civil/Social benefit, (3) Military activities, and (4) Commercial enterprise. The space economy resides at the intersections of space exploration, social benefit, defense, and commercial enterprise. It is the frontier's edge where companies, countries, entrepreneurs, and private investors are drawn by the promise of riches from high-risk and high-reward business ventures.

Most public awareness of space has appropriately been of the impressive scientific achievements, social impact, and military benefits of our national space efforts. Until now commercial space activities have been limited to reusable rockets, communications satellites, analysis of data harvested from remote

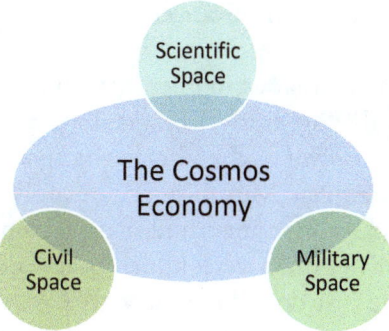

Fig. 4.1 Sectors of the Cosmos Economy. Traditional space activities include (1) military interests by sovereign nation states, (2) scientific exploration such as space telescopes and probes to celestial bodies in our Solar System, (3) civil space activities such as GPS and weather tracking, and (4) commercial space ventures currently in LEO (low Earth orbit)

satellite sensors, and other low-Earth orbit (LEO) ventures. These commercial activities, as phenomenal as they are, have not attracted the same degree of public interest as the early spectacular NASA successes of space walks and moon landings. The image a data analyst hunched over columns of numbers on a spreadsheet just doesn't generate the same excitement as Alan Shepard's unauthorized golf shot on the Moon.

But that will soon change; we are on the threshold of the new space frontier where the private commercial space sector is gearing up to change the traditional rules of commerce with business-related activities in space will soon redefine public perception of scientific, civil space, and even military space activities. Setting up shop to manufacture goods in zero gravity, to mine asteroids for high-demand rare materials, to produce food to feed a planet in vast space plantations, or to conduct special tourist experiences on the Moon is a big undertaking. The allure of a big payday outweighs the considerable risk of being an early entrant in the space economy.

Most importantly, space is no longer an exclusive U.S. domain. For ambitious nations, like China, who readily mix private and public investment, the attraction of space industrialization is obvious. The world's second largest economy is on a ravenous hunt for additional resources to satisfy their need for constant economic growth and to expand their international power base. Given the opportunity to establish a fresh new market instead of competing in the crowded and highly regulated markets on Earth, Chinese political and business strategists see the untapped greenfield of space as a blue ocean investment with tremendous upside. They are not alone. The next decades will see increasing investment in new space-based business ventures by private corporations, well-funded adventurous entrepreneurs, and aggressive sovereign states.

It's not hard to imagine that as momentum builds in the space economy other industrial firms that are not yet branded as space companies will find financial justification to pivot and embrace the space economy. Firms like Bayer (Monsanto), Caterpillar, Mitsubishi, Aramco, Foxconn, and even firms with a consumer product base may find a new home in space. As James Causey, Executive Director of SpaceCom, famously said, "every company is a space company—some just don't know it yet!" (November 20, 2019). Yet, how will entrepreneurs and corporate strategists know when to jump on the space bandwagon? Stay tuned.

References

Causey, J. (2019, November 20). *Every company is a space company, some just don't know it yet! (Op-Ed)*. Space.com. Retrieved from https://www.space.com/every-company-is-a-space-company-spacecom-op-ed.html

Lindsey, B. (2011, April 1). *Frontier economics: Why entrepreneurial capitalism is needed now more than ever*. Papers.ssrn.com. Retrieved from https://papers.ssrn.com/sol3/papers.cfm?abstract_id=1809996

Peck, J. M.. (1837). *A new guide for emigrants to the west, containing sketches of Michigan, Ohio, Indiana, Illinois, Missouri, Arkansas, with the territory of Wisconsin, and the adjacent parts*. 2nd ed. Boston, MA.

Sandburg, C. (2013). Chicago poems. Hardpress Publishing. (Original work published 1916).

Turner, F. J. (1893). *Wisconsin historical society the significance of the frontier in American history*. Retrieved from http://nationalhumanitiescenter.org/pds/gilded/empire/text1/turner.pdf

Turner, F. J. (2013). *The frontier in American history*. Createspace Independent Publishing Platform. (Original work published 1893).

5

Coming into the Cosmos

Today when you bring up the topic of space exploration, the notion of humans like us going to space to see what's there or to live and work, you are no longer considered foolish. But it didn't used to be that way. Today when you consider making an investment in a new space venture you will find like-minded venture capital firms and co-investors to join you. But it didn't used to be that way. Today space is regularly in the news with stories of private firms taking the lead in space tourism, of economical and reusable rockets, of the establishment of the Space Force as a new military branch, and of plans to colonize Mars by mid-century. But it didn't used to be that way.

It used to be that the mention human space travel and permanent extraterrestrial colonies would likely evoke rolled eyes and condescending comments. It used to be that your passion for human emigration to space might brand you as someone who lived in the fantasy world of science fiction instead of the real world of hard science. It used to be that business leaders, entrepreneurs, visionary thinkers, and ordinary people seeking a career that would make a difference had to keep their dreams of space to themselves.

But it's not that way anymore. Now space is front and center on the world stage and firmly accepted as the inevitable next step of humanity by everyday people and global leaders of nations and industries. That's the way it is today.

There are many reasons for increased attention about space settlement. At the top of the list is a general notion that our planet may be on a collision course with self-destruction. In this view it is widely believed that we are at the cusp of an historical inflection point; our planet as we've come to know it is changing and the pending outcome is likely not very good for people. In acknowledgment of the accelerating impact of human industrial activities on the Earth's delicate climate balance 195 nations have agreed to come together in the Paris Accord to pledge toxic emission reductions with hopes of healing the Earth. Only Syria, Nicaragua, and the U.S. have balked at participating in this rare show of multi-national unity.

Those who see the deterioration of Earth as a good reason to start planning for an alternative place for people to live point to these arguments:

- Climate change is altering our ability to produce food, live comfortably, and simply survive.
- Human impact on our planet is causing dramatic species extinctions, the death of vital eco-systems like the Amazonian rainforest, and the erosion of polar ice caps and glaciers thus affecting oceans and weather patterns.
- Industrial growth has put greater stress on our limited natural resources.
- The mantra of economic growth has created a zero-sum (win-lose) setting where income disparity is increasing with "have-nots" finding it harder to access clean air, unpolluted water, food, fuel, and resources for basic survival.

World citizens and business Investors see the mixed bag of visions for commercial space settlement as a potential opportunity to reverse out impact on Earth. These space advocates see space as an industrial safety valve, a cure-all solution, where essential toxic practices can be exported to space and removed from their destructive impact on our planet. Of course, there are other motivations for settling space. Some altruistic, some avaricious. But whatever the motivation there is a convergence of interests that are accelerating investments in the new space frontier.

People are motivated to make change happen for different reasons. Marketing folks and others interested in how people make decisions tend to think about motivation in different ways. Abraham Maslow comes immediately to mind. Professor Maslow's research summarized in his seminal *A Theory of Human Motivation* (1943) holds that people can be motivated by their hierarchy of needs. Those at the bottom of the stack are motivated to serve their personal needs of utility, shelter, food, and basic survival. At the very top of the pyramid are those few who are free of want or worry and who are not motivated by acquiring wealth but whose actions are driven by their need for self-actualization and spirituality. Maslow also considers the needs of people who reside at the various levels in between the bottom and the top of the pile—each level is best motivated by the context of their life, by how they best fit in to their social and economic world. Figure out where a person ranks on the ladder and you have the key to motivating them.

So, it is with big decisions like giving up a secure and comfortable life on Earth for a dangerous and unpredictable existence in hostile space. To make such decisions, it's helpful to understand the push/pull that drives motivation.

Push Pull. Embracing your Inner Cat

Does a person feel an emotional push that forces them to leave their current situation or does she feel an overwhelming pull to seek something new? This push/pull can be tricky. Some people are motivated when current circumstances *push* them out of their comfort zone while others are *pulled* to the prospect of something new that aligns to their notion of personal success and happiness. If someone is dissatisfied with their life or fearful about the future, they may feel a strong push to make a change. The push side of the decision focuses on leaving the present set of circumstances in a person's life, the pull is about being drawn to a new future life.

Let's assume you are an ordinary cat doing your daily explore of the neighborhood one sunlit morning. Your most favorite thing to do is to hunt for unsuspecting field mice. Your second favorite thing to do is to visit (i.e., taunt and tease) the neighbor's dog, Fritz, who is always tied up. But today Fritz isn't tethered. So, you are forced to run like hell to get away. It's a dream come true for unhitched Fritz and a nightmare for you. You escape (using one of your nine lives) by outsmarting Fritz. Next, you amuse yourself by chasing mice. Now the tables are turned; it is a nightmare for Rickey the rodent, but a dream come true for you.

Later that afternoon, as you get ready for a catnap in the safety of a sunbeam on the living room floor, you reflect on the morning's adventures and consider your inner catness; you wonder if you are a fear ridden cat chased and pushed by aggressive enemies or if you are a fearless hunter living your life in courage, pulled by what makes you happy? Put another way, do you live a life pushed by outside forces (Fritz the dog) that make your life miserable or are you pulled to actions that give you joy (Rickey the rodent)? Are you pushed to react with fear or pulled by the satisfaction that comes with courage?

(Note: I chose the image of a cat because, unlike dogs, I've never seen a cat chase its own tail. A dog chasing its tail is trapped in a ceaseless loop of pushing and pulling at the same time and always ends right back at the beginning again.)

Push Pull

Many see the prospect of space settlement as a refuge from our troubled planet. An escape. They may feel pushed to leave because they are anxious about the future of Earth. Constant wars, rapidly increasing population, pollution, depletion of natural resources, social unrest all build the case to escape Earth for those individuals who feel they would be better off someplace else, like space. They feel pushed away from their home. Others are drawn by the pull of something special in space, are less motivated by leaving, and are more motivated by the pull of a new life in the frontier of space.

One reason to leave is the belief that our planet is on a collision course with extinction. The prospect of settling in space is a chance to avoid the calamity of a sick world and to build a new home from scratch. Investors and adventurers who see space as the next big financial prospect want to make use of every opportunity to cash in on building a new world.

Coming into the Cosmos …

I wanted to know why some people feel they are pulled into being part of the space settlement movement. I also wanted to learn the reasons that others feel pushed away from living their life here on Earth. My conversations with space thought leaders revealed some major themes that is both good news and bad news about the emerging space economy.

Let's start with the bad news:

First, in case you're thinking about ordering your new bespoke spacesuit online, don't hold your breath. There will be very few people living or working space for the foreseeable future. In fact, it is unlikely that life-sustaining production of food, water, and shelter will make it safe and hospitable for large numbers of people to live and work in space for at a generation or more. Beyond that there is no compelling reason for people to go to space except as tourists or as refugees from some Earth-based disaster that will make the prospect of living in the stark dangers of space a better alternative.

The space frontier will be dominated by nations and large corporations leaving little room for individuals to prospect and strike it rich. Initial demand for human labor may be in the form of people able to repair malfunctioning robots or control their activities from nearby outposts.

Second, the first generation of the space economy will have some similarities with the American Western frontier of the nineteenth century. There will be lots of business opportunities, investments from large corporations, competition for territory and resources, and the lure of instant wealth for intrepid space adventurers. There will be little law enforcement, human rights will have little or no protections, and our current western notions of social justice and fairness will be sacrificed to commercial expedience. So, if you do find yourself among the few who are living and working in space and if you get in a jam, don't expect the local sheriff (or the cavalry) to come to your rescue. The prevailing rule of law in space for the first phase or two of economic adoption will probably be written in pencil by the company (or country) you work for—if it's considered at all.

Third, conducting business in the space frontier will not be easy. In fact, it will be damn hard and damn costly. There will be no ready-made infrastructure in space to enable resource extraction, mineral processing and production, manufacturing activities, transportation of goods, or any other element of a mature integrated economy we take for granted here on our home planet. However, this lack of infrastructure means there are opportunities for entrepreneurs to fill this vacuum. Revenues generated from transportation, energy, communications, habitat essentials (e.g., shelter, energy, food, water, security, etc.) are likely to represent the first phase, the frontier phase, of the new space economy.

Fourth, early entrants will face high barriers to entry; there will be lots of startup costs with few benefits in the short term. Those planning to enter the market in the earlier stages of development will need deep pockets to sustain their business plans. However, the advantage of being an early mover will increase the probability of a pay off in the long term. This means that those considering entry in this arena should not expect the traditional short-term return on investments as is the current thinking in corporate America. CEOs of firms aiming to enter the space economy should tell their boards not to expect quarterly accountability statements for quite some time.

Likewise, VC funds that typically expect investment returns in 5–7 years may have to adjust ROI expectations for a longer term. Space may be a target of opportunity for investors who are willing to stay the course for an extensive stretch of time.

Fifth, expansion of the new space economy will likely alter the center of gravity (CG) of the Earth's international economic structure. As the space sector moves from frontier market to mature market the impact of space industry on Earth's economy will be substantial. For example, an accelerated growth rate of space sector may shift the deployment of capital and possibly incur or magnify financial fluctuations on Earth. Whether or not the successes of space enterprises will cause this shift in the decades ahead, it is safe to predict that the impact of the expansion of the new space economy will significantly affect traditional earth-based enterprises.

Unknown, is whether Earth or space-based markets will be the dominant source of production and which will be the primary source of consumption. If the current technical limitations that restrict the amount of goods shipped from space producers to Earth consumers is not resolved, then the space economy will develop more independently and will likely adjust the types of manufactured goods, refined raw materials, and various services for space-based customers instead of gearing output for Earth-based consumers. The anticipated beneficial impact of space industrialization, where it is envisioned that resource-hungry, polluting, large-scale industrial manufacturing activities will replace those similar activities on Earth, will be severely mitigated by the current inability to ship unlimited tonnage from space to Earth port locations. Note that current ocean-going freighters carry container cargos weighing upwards of 10,000- to 15,000 tons of goods per ship. There is no such current capability for bringing similar loads to Earth.

However, if that technical hurdle is removed and large shipments can be regularly and easily transported down to Earth's surface, then the delicate balance of production and consumption will shift to follow the best source(s) of revenue. Note, we are currently experiencing a spate of spaceport construction all around the planet. These futuristic facilities are designed to accommodate the expected increase of space tourism with ancillary facilities for entertaining the expected glut of space travelers. No new spaceport designs have yet to house facilities for shipping or receiving cargo to and from space.

While this co-dependency of Earth and space industrial activities could be either good news or not, the immediate effect of a successful space economy on Earths precarious balance of trade could be potentially disruptive. Unknown at this time is who will be dependent on whom for economic sustainability. If an unmanaged imbalance of trade develops, then the impact on Earth investors who are dependent upon revenue from space-based industries could be severe.

Finally, there is no such thing as an all-encompassing master plan for developing space. Just as with the opening of the American frontier, it's each company (or nation) for themself. A fundamental tenet of traditional economics is that companies and individuals act in their own self-interest. Those ventures who are willing to share resources and find integrative opportunities to connect their business with other complimentary firms will have a greater chance of generating financial success than companies who insist on doing things all on their own. To be sure, the resurrection of the National Space Council under the direction of the White House is a step in the right direction to create a common sense of purpose about the future of space.

Partnering major aerospace prime contractors with government agencies to create a common vision and roadmap of the future is a positive step on the way to creating a shared master plan. But the nature of competition is such that, given the opportunity to innovate or stake a flag in a new frontier, independent firms are likely to view any such master plan as merely a suggestion.

Now the good news.

For companies with the ability to make a long-term investment the allure of establishing and defining a new market will yield significant returns as the space economic sector transits from its nascent frontier market beginnings to maturity. Successful early entrants will have already struggled through the steep learning curve of operating in space and will be able to leverage their knowledge to partner with, or vend to, later entrants. These innovators and risk-takers may yield more than short-term profit. Early adopters will be well positioned to create a long-term economically sustainable competitive advantage through deployment of leading-edge technologies and the first-come-first-served benefits of moving quickly to exploit the opportunities of the young space frontier.

The second advantage of the dynamic nature of the space economy is that early movers in the beginning phases of adoption will be positioned to provide invaluable support services and infrastructure to those who follow in the later phases. Those who move into this new arena during the early phases of adoption will likely learn from trial and error about how to maneuver, how to change course and pivot where necessary, and how to succeed and prosper in space. For example, space agri-business (e.g., fiber, foodstuffs) is a later phase enterprise that of necessity will rely upon transport providers (an early phase enterprise) to have developed the capability of carrying cargo to market. Likewise, early outposts will be well positioned to serve other commercial activities with their "rest stop" model of providing fuel, housing, storage depots, entertainment, and other frontier necessities.

Early entrants will benefit from the Principle of Cumulative Advantage which says that small advantages gained by being an early risk-taker can compound over time and grow to become an increasingly bigger market advantage. This well-known effect is captured in the famous phrase, "the rich get richer and the poor get poorer." This is like an investor who starts saving money early in his career versus someone else who starts to save or invest much closer to retirement. The early entrant gains the advantage of compound interest over a long time and ends up with a healthier portfolio than the laggard.

The third benefit is the lack of enforceable regulation on the space frontier. The unformed social ecology of space will afford risk-takers the opportunity to exploit mineral resources and develop efficient processing capabilities with minimal regulatory oversight and thus enable them to be providers to manufacturers and others who will follow them to enter in a later phase of the space economy. Early mainstream through late adopter phases will depend upon the commercial output of earlier adopters and innovators who came to space before them. Being an early mover will be a distinct advantage for those who choose to take the risks associated with this strategy.

The fourth benefit leverages the anticipated wave of technological innovation that will come from doing business in the unusual setting of space. New commercial production techniques and technologies will undoubtedly be developed for space that will likely eclipse those in use back on Earth. Reverse migration, from space to Earth, of new technologies developed for space for use on Earth will upend traditional methods and practices. This likely will apply to space-based manufacturing, energy acquisition, infrastructure, food production, transportation, housing, and other commercial activities that will quickly be adopted into Earth-based business operations. The acceleration and adoption of these new techniques and technologies by traditional Earth-based industries will likely result in increases in productivity, competitiveness, and revenue. However, as with every innovation milestone, the introduction of niche technologies into mainstream industrial practice may create the unanticipated effect of reducing the need for lower-level jobs (and maybe high-level jobs, as well). Another blow for the human labor force in age of space industrialization.

The fifth benefit of being an early adopter is the impact of space itself on creating new opportunities for firms seeking profits in space. These opportunities may take the form of acquiring and utilizing resources in space, leveraging the zero-gravity environment for manufacturing, harvesting ice or water for energy and agri-business, or a variety of other benefits unique to space that make doing business in space a competitive advantage. There is broad enthusiasm for the potential profits from mining asteroids or other deep space bodies for raw materials. Likewise, zero- or micro-gravity manufacturing presents new opportunities for a variety of innovative enterprises.

Businesses operating in space will also be operationally different from traditional businesses back on Earth. The high demand and high revenue potential of using the abundant ice or water available on the moon, Mars, and other places in space coupled with limitless capacity to grow and process foodstuffs and other agricultural products, will create tremendous opportunities for space agri-providers. Space agri-business, in a variety of forms that include grain crops, meatless protein products, and industrial fibers like cotton, flax, or hemp, may prove to be a significant industry in the space economy that will have an immediate effect upon markets on Earth.

The sixth advantage is the low demand and high cost of human labor in a setting more compatible with artificial intelligence and robotics. The economic advantage of operating an enterprise with lots of robots and very few people is obvious. The more robots in use, the greater the savings on labor costs and benefits burden of a traditional human workforce. The result is that enterprises in space will benefit by severely reducing or eliminating labor expenses. Human labor costs are notoriously very high because of all the salary burden costs that include health insurance, taxes, safety accommodations, pensions, and other add-ons. The indirect overhead expenses of furnishing livable workspace, safety concerns, and other expenses make human labor in space prohibitive. It's no wonder that employers would prefer robots over human workers. Further, humans tend to work between 40 and 60 h a week on average whereas robots can work nonstop all 168 h in a week (7 days × 24 h) and don't need special housing, food, air, bathrooms, entertainment, vacation time off, paternity leave, a corner office with windows, or other ancillary human-centric expensive enhancements. Providers of robotics and artificial intelligence capabilities, take note.

Most significantly, the scope and priority of a human-centric space mission is fundamentally different from the same mission supported solely with automation. Once humans are introduced into the process, the goal of the mission dramatically shifts away from merely accomplishing a technical mission to maintaining the safety and health of the human with all the costs associated with supporting this revised objective. If people are taken out of the cost equation (and replaced with robots and automation) then the mission reverts to simply accomplishing the task at hand. People come at inflated costs in space and these costs include shifting the goal from a straightforward set of business objectives to securing the safety and well-being of the human crew.

Finally, early entrants who gain market advantage will find themselves in a better position to exit. An exit strategy is the process by an owner to transition out of the enterprise. This is usually accomplished either by direct sale of the business to another firm or transfer of ownership through a merger with another company.

References

Maslow, A. H. (1943). *A theory of human motivation*. Eastford: Martino Publishing.

6

No Country for Earthmen

Much is still unclear about the long-term effect of space environment(s) on humans. Concerns about the impact of prolonged social isolation on the mental stability of space explorers have been studied as well as the long-term consequence of confinement. Some studies have looked at samples of submarine sailors and at prisoners in solitary confinement (Haney 2012). In the case of submariners, according to Dr. Henry Schwartz, M.D. (Captain USN, Ret.), they are vetted and trained for the ordeal of months in close confinement with others. They also are kept busy throughout their day with clearly defined accountable tasks that allow them to maintain a sense of professional crew identity in the group as well as a sense common purpose and accomplishment.

This contrasts with the long-term effects of prison solitary confinement. In one study of prisoners it was found that an isolated prisoner may be able to manage the ordeal but are likely to become acclimated to the severe isolated environment and then prefer living without regular human contact. The principal investigator of the study, Dr. Craig Haney (2012), observed that, "One of the very serious psychological consequences of solitary confinement is that it renders many people incapable of living anywhere else and when released [they] are often overwhelmed with anxiety." A study by NASA (The Human Body in Space 2011) suggests that no matter how well selected or trained the space crew member may be, he or she may become attuned to the severe social isolation of space and may need to manage the transition to live among others in a more complex social setting.

Submarine isolation or prison solitary confinement aside, the ordeal of space travel provides additional threats to our species. Deadly radiation, bombardment by micro-meteorites, and the challenge of living in an artificial environment like a spacecraft or an outpost habitat top the list of lethal threats. Early space explorers on a seven-month sojourn to Mars will likely have to make do and live off the land for several years before their return trip to Earth. To facilitate this, initial Martian expeditions will rely upon staging food, fuel, and other vital goods on Mars or in Martian orbit well in advance of human arrival. The hostile environment of Mars will threaten the fragility of their bodies and push their psychological stability to the limit.

Once out of the relative safety and familiar comfort of the Earth's gravity and atmosphere the human body is under constant stress. A short list of the physical effects caused by being in space includes:

- Humans lose 1–3% of their bone mass *every month* spent in space.
- The immune system deteriorates and loses its effectiveness.
- The cardiovascular system gets deconditioned.
- The neural vestibular system has significant challenges.
- A micro-gravity environment can permanently affect eyesight.
- Muscle tone can deteriorate.
- Ambient space radiation may alter human DNA.
- Fungal spores can survive intense space radiation and potentially pose residual health risks.
- Spaceflight reduces heart rate, lowers arterial pressure and increases cardiac output.
- Cancer is a big risk during prolonged spaceflight because of exposure to radiation.
- Spaceflight creates a stressful environment, awakening dormant viruses.
- Four of the eight human herpes viruses were detected, including oral and genital herpes, chickenpox, and shingles.
- With increased confinement comes increased behavioral or cognitive problems and psychiatric disorders.
- Severe stress due to no fresh donuts!

On a more positive note, as noted by NASA administrator Jim Bridenstine (Williams 2019), the micro-gravity of space creates research opportunities that may solve some of our species' inherent weaknesses. Some of the research conducted in space takes advantage of the micro-gravity environment to develop new pharmaceuticals or to learn how to 3D-print human organs using adult stem cells. But there is also a downside. Strickland (2019) reports that spaceflight activates herpes in astronauts.

Setting the stage for future generations of humans in space assumes two important things: (a) pregnancy and (b) giving birth in space to enable those future generations. So far, neither of these milestones have happened. At this writing over 60 women have gone to space; none of whom were pregnant or gave birth while in zero gravity. Without Earth's gravity birth in space might be more difficult for the expectant mother to push her baby out of the birth canal. Another complication is that bone mass is reduced 1–3% for each month spent in space. Loss of bone density could cause the mother's pelvis to fracture which might lead to more Caesarian births than natural births in space. This may lead to changes in the physical structure of people born in reduced gravity like larger heads and change in skin color due to the lack of Earth's protection from ultraviolet sunlight. There is some thought that future generations of people born in space may evolve to be physically different from today's humans and may be eventually considered as a different (sub)species. We will carry our human evolution with us to space even though it will be several centuries before we learn the outcome of our evolutionary development. Among many concerns about human biology in space is the idea that babies born in space may have much larger "alien-like" heads (Kim and Wilkin 2019).

Frankenauts

Human survival in space may depend on the possibility, the probable likelihood, of altering the human genome to make the spacefarer more adaptable to the harsh space environment. My first readings of these articles left me with little assurance that this topic should be included in this book project—after all, I thought, this topic smacks of science fiction. But the topic nagged at the back of my mind to the point that I am including a brief reference here for two reasons. First, the science that supports altering the human makeup seems to be both credible and doable. Unlike time travel, faster-than-light warp drives, or teleportation, the ability to enhance a person's corporal ability to withstand more radiation and put up with other life-threatening distresses may well come to pass. And second, I had to discard my concern about including a topic that is on the cusp of science fiction because, when push comes to shove, practically everything in this book, no matter the level of research or depth of interviews with experts, is really about the future; conjectured on the cusp of science fiction and historical fact.

When two pieces of a puzzle don't fit neatly together there are limited tactical solutions. Either (a) give up the goal of ever solving the riddle, or else (b) change the dimensions of at least one of the pieces to force a solution. When it comes to the question of how humans can live comfortably in the unfriendly environment of space, the former option (giving up) is out of the question; there is too much forward momentum, too much fiscal capital and political capital already spent to make the puzzle parts fit. It falls to the latter option (changing how to make the pieces fit) to find an answer. In this case the puzzle is how to find a way for humans to easily inhabit the deadly space environment. The mystery puzzle piece is human bioengineering.

Up until now most of the effort in this pursuit has been technical adjustments to the space traveler's environment. Examples include designing habitats that thwart gamma ray bombardments, providing devices that make space water drinkable, designing vitamin supplements to counteract the deficiencies caused by prolonged space travel, rigorous exercise regimes to maintain bone mass otherwise lost in zero G, sanitizing the interior of a spacecraft to minimize growth of fungi and other microbial stowaways, providing nutritious food supplements to maintain overall health. Not to mention the careful selection and extensive training designed to acclimate the space traveler to what they might expect on a long voyage to a distant planet or deep space outpost.

But there are some researchers who want to explore a different tactic. They don't want to bother with only altering the environment of the spaceship. They also want to alter the makeup of the spaceship's occupants.

It's one thing to have astronauts who are engineers and quite another to biologically engineer the astronauts to alter their genetic structure to make them more adaptable to space radiation and other mortal threats. This implies creating a "spaceman" subspecies that would be fundamentally different from Earth men. How different is yet to be determined through strategies like epigenetic engineering, a method that allows for switching individual genes off or on.

Another approach is even more drastic. By combining human DNA with the DNA of another species, like tardigrades who appear to be able to survive space unaided by life supporting air or water, it is suggested that future space folk would be more resistant to radiation and other benefits (Bittel 2016). Tardigrades are microscopic bug-like animals that look like a cross between a caterpillar and a nude mole rat. These aquatic invertebrates are known for their capability to withstand a broad range of extremes, including high and low temperatures, lack of nutrients, near total dehydration and the stressful environment of space. If I was looking for a date, then a tardigrade wouldn't even be at the bottom of my list. If I was shopping for a match for my DNA to help me survive in space, then I might be more open to an introduction.

Genetically engineering human DNA to make people more compatible with space would force evolution in a new direction. It is assumed that humans will adapt and evolve to some new altered version of Homo Sapiens after living in space for generations. But some researchers want to make that happen sooner rather than later. Space survivability, especially for long-haul trips to Mars and beyond, may come at the high price of human genetic reengineering (Gohd 2019). There are reported experiments on the ISS using CRISPR gene-altering technologies to explore how this may be implemented (Houser 2019). Human evolution resulting from living in space isn't a matter of "if" but, rather, a matter of "how."

A variant on the theme of bioengineering the human genome to create a new human space "Frankenaut" is the Lego-type approach—creating something new from a collection of biologic building blocks. In this version of re-forming the puzzle piece in order to solve the riddle 3D bio printers are used to manufacture replacement organs and other body parts to either repair damage caused by extensive time spent in space or to swap existing healthy human parts with upgraded replacement parts made to prevent and mitigate potential damage caused by the destructive effects of the space environment. If you think it annoying when your computer tells you it needs to download a newer version the operating system, then imagine what it would be like if one or more of your body parts alerted you about needing a quick download of their latest version of *"Heart 2.0"* or *"Kidney 3.5.1"* so you could keep living. Of significance is the potential for reverse migration of biotech from utilitarian space applications to Earth consumer markets. Current mass marketing of Lasik eye surgery and dental implants would pale compared to future a future industry created by medical practitioners peddling a complete overhaul of replacement and upgraded body parts that will help to enhance health, reduce the effects of disease, and prolong life. The impact on the current traditional health care industry will be enormous.

This is not a brand-new idea. Classical science fiction literature has already deeply mined the rich potential of this narrative vein. Prof. Lisa Yaszek of Georgia Tech, a leading researcher of science fiction literature, suggests these well-known examples of stories dependent upon the element of adding/changing/removing body parts:

- Mary Shelly's *Frankenstein* (1818), an allegory of the new industrial age and the collision of science with sentient humanity, first proposed that life could be created by (re)engineering thus turning scientists into gods.
- *The Strange Case of Dr. Jekyll and Mr. Hyde* (1886) by Robert Louis Stevenson explores how hubris and scientific experimentation can expose the dark duality of human nature with disastrous consequences.

- *The Island of Doctor Moreau* (1896) by H.G. Wells accounts the horrors resulting from a mad scientist's godlike vivisection experiments changing wild animals into civilized men and control the direction of evolution.
- *Metropolis* (1927), a visually stunning film directed by Fritz Lang that employs the creation and then transformation of a maniacal robot into a false savior of the working people.
- *This Island Earth* (1955), a mid-century Sci-Fi thriller where aliens and their Earth counterparts must undergo physical transformation to survive in each other's otherwise deadly world.
- *The Fly* (1958), an experiment of molecular transportation goes wrong when a scientist attempts to transport himself. A fly enters his transport chamber resulting in the combination of the Fly's DNA with the scientist. After that, things don't go well for either the scientist or the fly.
- *Blade Runner* (1982), Ridley Scott's dystopian film about seeking and destroying bioengineered *replicants*, synthetic humans created to work at space colonies, who have fled back to Earth.
- *The Six Million Dollar Man* (1973–1978), a popular TV show predicated on the transformation of a normal man into a modern-day superman by extensive bioengineered upgrades to his body.

Consider that these fictional accounts do not end well for neither the hybrid creatures nor for their creators. But there are medical examples from real life that attest to a more positive outcome of human physical enhancement. Today there is a range of body-altering medical procedures that reduce suffering and prolong life. Heart transplants, dialysis treatments, medications to cure disease, kidney and other organ donations, pacemakers, the creation and use of artificial skin, cochlear implants, Lasik eye surgery, prosthetic limbs, artificially grown skin for burn victims, and an increasing number of new medical interventions designed to alter and improve the human condition. It is easy to imagine that these innovative remedies will continue to increase.

References

Bittel, J. (2016). Tardigrade protein helps human DNA withstand radiation. *Nature*. https://doi.org/10.1038/nature.2016.20648.

Gohd, C. (2019, November 7). *Can we genetically engineer humans to survive missions to Mars?* Space.com; Space. Retrieved from https://www.space.com/genetically-engineer-astronauts-missions-mars-protect-radiation.html

Haney, C. (2012, October). *Risks of solitary confinement*. https://www.apa.org. Retrieved from https://www.apa.org/monitor/2012/10/solitary

Houser, K. (2019, May 28). Astronauts use CRISPR in space for the first time ever. *Futurism*. Retrieved from https://futurism.com/the-byte/iss-astronauts-crispr-space

Kim, G., & Wilkin, R. (2019, July 23). *If humans gave birth in space, babies would have giant, alien-shaped heads*. Business Insider. Retrieved from https://www.businessinsider.com/humans-gave-birth-space-earth-giant-alien-heads-2019-7

NASA. (2011). *The human body in space*. Retrieved from https://www.nasa.gov/hrp/bodyinspace

Strickland, A. (2019, March 20). *Spaceflight is activating herpes in astronauts*. CNN. Retrieved from https://www.cnn.com/2019/03/18/health/astronauts-spaceflight-herpes-study/index.html

Williams, H. (2019, January 3). NASA administrator Jim Bridenstine on life off earth. *Wall Street Journal*. Retrieved from https://www.wsj.com/articles/nasa-administrator-jim-bridenstine-on-life-off-earth-11546530870

7

Turning over [Space] Rocks …

Assumptions About Space Business

Until now, interest in the *business* side of space has been mostly in the boardrooms of major corporations and in the workshops, hangars, or garages of the innovators and entrepreneurs who are inventing the future. No matter the size of the firm, most space organizations accept that the funding model is changing. Well-established traditional aerospace companies and forward-thinking nations across the planet have come to recognize that the old financing platform is shifting away from public funding and moving to private equity and VCs for underwriting new space ventures.

The most direct path to space is via great technologic developments, but the technology that enables space is only a supporting character in this story; the heroes of this story are business leaders who see the potential of space in their company's future. Their driving motivation is profit in service to a grand vision of our future in space. To pursue this story, it is best to not simply focus technology but to acknowledge the business-minded risk-takers and entrepreneurs, the financial adventurers, who will enable the next step in Earth's economic development.

There are those who look to human activities in space as a solution, a palliative, for the ills of our problematic planet. But building a space economy is not the easy quick fix to all of today's economic, environmental, or social problems. Those who believe that space will somehow offer a utopian solution to the misdeeds and missteps of our species are overly optimistic; the same folks (us) who fouled up Earth with social problems, pollution, bureaucratic formalization, and ego-driven political shenanigans are likely to replicate the same mistakes on Mars or wherever people plan to go in hopes of escaping our present crisis-laden world. This is not a story about how to build Shangri-La—it is a story of how to build civilized commercial space.

Sorting Through the Assumptions

When space professionals gather to chat about space there is a menu of given assumptions that everyone takes for granted. In my conversations with a cross-section of space business leaders and academics I started to see common threads and trends. It soon became clear that some of these beliefs are probably true, others fall into the realm of magical/ wishful thinking, and then there are those iron-clad beliefs that are probably dead wrong. Some of the assumptions that emerged—good, bad, and ugly—include:

(a) Some common assumptions I believe to be true in the next 50–100 years.

- Human settlement in space is a sure thing and will happen in the twenty-first century.
- Military activities will be a leading motivation for developing space.
- An alternative habitat for Earth's citizens will be a leading motivation for developing space.
- Colonial space settlement will develop in accord with twenty-first century *national* initiatives.
- Industrial space outposts will develop in accord with individual *corporate* interests.
- Just in case Earth is doomed (because of likely inept human management) it is important to build "lifeboat" settlements in space. This is a prevailing assumption but assumes a zero-sum solution to our planet's climate problems.

(b) Assumptions I believe are unlikely in the long run.

- Earth will always play the dominant role in the space economy. Nope.
- Space settlements will develop uniformly across the solar system. Sorry, human habitation of space will be organic and chaotic.
- Space policy will set an enforceable standard for law-abiding behavior in space.
- The development of space commerce will rely primarily upon the extraction, refinement, and use of raw materials found in space. There

needs to be a sustainable market for all the raw materials—this will take some time to develop.
– Scientific activities will be a leading motivation for developing space. Scientific inquiry mostly relies on government funding. The new space economy is about private investment.

(c) Assumptions that depend upon the yet-unsolved technical problem of bringing large commercial payloads from space to Earth: the "Down-Mass" problem. Until then, the following assumptions fall into the "not yet" column.

– Large-scale industrial manufacturing in space will reduce pollution and mitigate the impact of environmental contamination on Earth.
– Growing and processing food on an industrial scale in space will help to alleviate hunger and mitigate the impact of environmental contamination on Earth.
– Earth will be the primary consumer of goods manufactured in space.

Outer Space Has a Down-to-Earth Problem—Getting Down with the Down-Mass

One of these assumptions is the lynchpin for many of the others, the assumption that we will have the technical ability to ship immense cargo tonnage originating from space to Earth's consumers. It turns out this assumption is wrong: no one has figured out how to bring all that freight safely back to Earth. This is known as the *down-mass problem* and it may represent a major speedbump on the road to free and open space commerce.

Our current launch capabilities enable us to ship lots of stuff up into space. The cost per kg keeps dropping due to improved technologies and cost management. But we have yet to develop the ability to load thousands (or even dozens) of cargo containers with goods made in space and ship those goods to Earth-based markets. Picture a humongous freighter piled high with thousands of tons of consumer electronics, cars, T-shirts, cat food, and whatever else American consumers crave. Add at least a dozen more container ships just like it moored at the Port of Long Beach in Southern California every day. Next, multiply this growing cargo quotient by the dozens of ports around the world that manage the transshipment of goods from manufacturer to wholesaler to retail store and finally to the individual consumer. That's a lot of potential freight moving to markets all over the planet all the time. Because we don't have the down-to-Earth technology to ship freight from space to Earth the promise of the multi-trillion-dollar space economy is likely to stall.

I was surprised to learn of this critical limitation. I just assumed that it was easier to ship stuff from space to Earth (it has to do with that whole gravity thing). But I assumed wrong. I realized that the implication of this limitation is more than just a logistical challenge: it is the critical variable, the fork in the road, that will define the future direction of the new space economy for decades to come.

Significantly, until this critical technical glitch is solved, the vector of the space economy will aim away from Earth and point instead to serving budding markets in the new off-Earth space-based economy. This shift will cause the space economy to adjust itself to best serve itself. In this setting the products and services that will be manufactured in space will be designed to satisfy the specific demands of space-based customers and consumers instead of targeting the broader (and more profitable) consumer base on Earth. Not only will this impede the rate of growth of space commerce but if Earth isn't the primary consumer of space output, then it means that Earth will not play a central role in the new space economy and will assume a lessor supporting relationship with space commercial activities. All those planners and strategists who assume that Earth will naturally play a central and controlling role in the space will be disappointed. So, when I would raise my hand and ask an expert how the down-mass problem would be resolved and who was working on the technology I would get the typical answer: "Gee I don't know of anyone who is working on that problem, but I'm sure someone must be." I'm still waiting for an answer.

One of the well-worn tropes used to describe the development of space is the building of the Transcontinental railroads and their impact on the economic development of the western territories in the late nineteenth century. This is an overly convenient analogy. Establishing rail access from coast to coast certainly did create opportunities to ship goods, pioneers, soldiers, infectious disease, livestock, and raw materials in larger quantities and in less time than ever before. It is true that wherever the rails went towns sprung up to connect their local commerce to the economy of the greater nation, and the rest of the world. One of the major differences between railroads and rockets is that only one generates revenue via round trips (railroads) unlike the other transport vehicle (rockets) which is limited to producing revenue by going in only one direction (up). This important difference is one reason the convenient analogous comparison of railroads with rockets is deficient. The other reason the railroad example won't fly in space has to do with the heavy dependence upon labor in the Olde West versus reliance on automation and technology for production in space. When you don't need lots of workers you don't need the transport to move them around or to bring them the necessities of life in space.

The funding model of the railroads, the Public-Private Partnership (PPP), is often held up as the leading solution to kick-start new ventures in a frontier economy where little or no substantial infrastructure exists. The PPP business model brings public government funding and private industrial expertise together as partners to solve a technical problem or to build a project that will benefit a broad base of citizens. NASA projects have employed this model for decades by funding private contractors who bid for the opportunity to provide products or services that meet the specifications and requirements of a proposal. One important caveat of the PPP is that profitability is not considered as a measure of success. In the nineteenth century five of the six transcontinental railroads were funded using the PPP model and none of them achieved profitability. Ever.

Just in case you were wondering if settlements in space are a "done deal," think again. There are arguments for not settling space. In fact, there are several arguments that challenge the current fever-pitch enthusiasm about migrating to Mars and establishing outposts in the far-flung corners of our solar system. As with most things concerning a subject that attracts ardent enthusiasts, many of the cautionary arguments opposed to rushing to space are logical, some are technical, but all the objections come with a strong dose of emotional fervor. Apart from the substantial warnings about the demise of our planet and the very real impact of climate change on species extinctions, including our own species, there are those who say, "Yes, but …".

Yes, the planet is changing, **but** apart from that reality are independent reasons to settle space. It is an opportunity for human expansion, a place with untold riches just ripe for development, and because of current technology the solar system is now simply an unexplored extension of our planet's back yard. Whether or not the Earth is in dire straits, this is the moment in time for human migration into space.

Yes, there are problems here on Earth caused by human industrial activities, **but** the reports of Earth's death are greatly exaggerated. The current problems we have with climate change such as melting ice caps, rising sea levels, boiling hot summers, deep freeze winters, massive fires caused by spotty rainfall are all real. But they are merely a natural cycle of nature. In time these catastrophic changes will either revert to the way things used to be or we will find ways to adapt to the new normal.

Many who feel this way maintain that reports of the imminent death of our planet are at best a misread of the data, and at worse a fictitious hyperbolic exaggeration. Those who see no urgency to settle space because of climate change claim:

1. There is still plenty of time to correct the wrongs committed in the name of industrial progress.
2. The whole topic of climate change is a myth or, at least, an exaggeration. Climate change is a hoax promoted by Luddites opposed to progress with an anti-business agenda.
3. There have always been catastrophic changes on our planet (ice ages, asteroid extinctions, pandemics, alcohol-free beer) so the specter of a changing climate is nothing new. Go with the flow.
4. The more conspiratorial among this group suspect that the entire topic of space settlement is being promoted by cynical investors and promoters who only want to get rich from developing space, who care little about creating a better world, and who persistently exaggerate the threat of climate change only to seed fear and worry.

Yes, the notion of settling space sounds like a reasonable idea, ***but*** why should we have to act now? This argument rejects a sense of urgency about space settlement. Like Scarlett O'Hara in Margaret Mitchell's *Gone with the Wind* (1936), there is plenty of time to act. After all, "tomorrow is another day." They hold that the exploitation and settlement of space does not need to happen soon, and because there are too many technical and financial hurdles the entire topic is best left for later—a lot later. Decades later. Lifetimes later. Even centuries later. While current settlement advocates notably speak about establishing outposts and settlements in space with a high sense of urgency and want to move the process along as fast as possible, this alternative group prefers a more sedate "wait and see" approach.

A second argument against space settlement is financial. This argument is based on the assumptions that (1) space activities are primarily funded by governments and (2) space activities are an exclusive American domain. Adherents to this perspective believe that tax monies spent on space activities yield little fungible return on the investment, that we've already put our flag on the moon, and that funds allocated for space could be better spent on other, more impactful, social activities. This transactional either/or approach often ignores the possibility that responsible social programs and other ventures like investment in space can happen concurrently. But times have changed because the way space is financed has changed. Most of the money invested in commercial space ventures (77.1%) comes from private (non-governmental, non-tax) sources with only 22.9% of space funding coming from government sources. Likewise, 61% of private space investment comes from non-US sources (Bryce 2020). There is every indication the trend of increased private space funding will continue to eclipse the investment by government resources (except for military spending).

The plea for more government funding of Earth-bound social programs is sincere and not to be dismissed. However, the case that funds should be deployed for socially responsible programs instead of for space becomes less relevant as the government role of funding space decreases. A need for government financial support remains. Private investors rightfully expect a return on their investment but not all space missions are commercial. There are purely scientific and extremely important space exploratory missions that don't attract investors and will continue to need government financial support. A major flaw of this argument is that it assumes a zero-sum model where monies are allocated for either project A or for project B. The alternative is that both (or neither) projects can be equally funded or can access funds appropriate to their project needs.

There are, of course, other legitimate reasons NOT to settle space. Anti-settlement proponents cite such issues as high cost and minimal return, the risk to human life in the unfriendly environment of space, inadequate or immature technology not yet ready to sustain off-world life, Earth is the only place in the cosmos where humans are meant to live. Even proponents of space settlement balk at the murky financial projections of return on investment, especially in the short term.

And finally, **yes**, there is a counter argument to space settlement that believes private enterprise is unlikely to sustain long-term funding of space economic growth. **But** there may be a kernel of truth in this assumption.

First, corporate enterprises are conditioned to think in short-term quarterly chunks. They issue earnings reports and hold investor guidance conference calls quarterly. This affects how corporate strategies are developed and executed and infuse the corporate culture with a short-term mind set. The creation of a sustainable space settlement economic base is a long-term commitment which may cause some investors a considerable degree of angst in time. As we have painfully learned, national and global economies tend to have dramatic ups and downs. In turn, these "adjustments" can have serious consequences in financial markets which may cause private investors of space ventures to rethink their funding strategies and possibly rebalance their portfolios to shed their space investments. This argument gains traction mostly because of the natural volatility of financial markets. The other side of this position is a mix of confidence and commitment. Most of today's expected VC returns are based on a five- to seven-year span; investors in venture funds expect a return on their funds in that time frame. The shorter the period the better. Creating a new market in space—a place currently without infrastructure or a measurable consumer base—will not happen in the short term. Space has a very long-term investment horizon.

One unanswered question is what would space settlement be without space settlers? When I asked this question at a large national space conference (National Space Society 2019), I was met with a mix of confusion along with steam-blasting-out-the-ears hostility. The reason for my question was to explore the degree that humans were necessary to create space-based commercial activities. I assumed that the alternative to millions of people would be tens of millions of robots and automated production that would take the place of traditional human labor and reduce the liability and risk of coddling a fragile and vulnerable human workforce in the highly risky platform of industrial space. But, instead of the thoughtful engaging response I hoped for, I was treated to a hostile (and condescending) diatribe about the definition of settlement; settlement, I was instructed, "is about *people* establishing themselves in space—not robots." That was my introduction to the strong religious-like zealotry that prevails when it comes to discussing the role of humans in space.

I think there is a reasonable alternative argument where mechanized space outposts and commercial centers could run quite well with few if any people. But on occasion, contrarian argument and logical discourse is eclipsed by the emotional "belief" that space settlement is the manifest destiny of humanity, not to be challenged by an unbeliever espousing a logical alternative.

So it goes.

Reference

Bryce Space and Technology. (2020). *Start-up space; update on investment in commercial space ventures.*

8

Planning on Purpose

Creating a new enterprise is no trivial enterprise, even more challenging is the process of turning that creative idea into reality (innovation), especially if the new enterprise is meant to be in space. It turns out that lots of folks are creative and have brilliant ideas about how to make the world better, how to be successful, and how to make the solar system a place for human settlement. Lots of these creative ideas are founded on sound science and apply laws of physics while some other ideas are best left to science fiction. In either case the owners of these creative notions believe resolutely in their efficacy; they believe that they have found a solution that will change the future and remove a critical barrier to human habitation of space.

Some of these solutions may likely to be of actual benefit while other ideas are merely elaborate solutions in search of a problem. Regardless of the utility of the creative idea the missing connection is whether the idea will serve a greater strategic plan. Ideas by themselves are not enough to bring about change. Ideas aligned to a greater plan have a much better chance of making a difference.

Much has been written about the importance of strategy and how tactical actions underpin a strategic operation. Classics about strategic planning from earlier eras like *On War* by Carl von Clausewitz's (1997), *The Art of War* by Sun Tzu (2010), and *The Prince* by Niccolo Machiavelli (2016) immediately come to mind. These, and other early treatises born from a regimented military mindset and a feudal approach to power, introduced the practice of careful deliberation before action instead of just responding to threats or confrontations. Strategic thinking inaugurated the novel notion that generals and political leaders should not act impetuously but should assess the current state, evaluate resources and capabilities, consider strengths and weaknesses of the enemy as well as themselves, create alternative scenarios, and only take action after a reasoned plan has been devised that weighs the likely outcome of all options. One result of adopting strategic thinking as a matter of good military or political practice is that the leaders who rose to positions of power were more likely to be the types of persons who would excel at the chess moves of strategy. Leaders in the new industrial age were better prepared to temper their thinking along more complex lines of reason. Their political and economic successes were the result of complex reasoning not required in earlier, simpler times.

Harvard Business School Professor Michael Porter formalized the modern strategic planning process with his classic book, *Competitive Strategy* (1980, 2004). His highly influential book presented five forces that were key in determining how a firm should evaluated their strategic context. These are the threat of new entrants, the threat of substitutes, the bargaining power of customers, the bargaining power of suppliers, and competitive rivalry. He is famously quoted saying, "The essence of strategy is choosing what not to do."

Why this Discussion of Strategy?

It is one thing to have a vision of the future; leaders need to create a vision, a narrative story, to establish the direction for their organization so that the managers will deploy resources and capital properly. The more specific and concrete the vision, the easier it is to achieve and to measure a manager's success. It is important to have an idea about where you want to go before you start a journey. As the old homily goes, "if you don't know where you're going, they you'll never know when you get there." Steven Covey put it best in his book, *Seven Habits of Highly Effective People* (1989, 2005) when he said, "start with the end in mind."

Wedged between grand vision and execution is strategy; the deliberate roadmap based on thoughtful consideration that delineates what *should* be done from what *might* be done.

Ready, Fire, Aim …

A more formal structure for creating a strategic plan of action can be borrowed from common business disciplines like executive coaching and project planning. The deceptively simple steps are:

1. **What is the current state?** Assessment of current state will yield a baseline understanding of the good and bad, the possible and the impossible, given the current operating environment. How will we know if we are successful?
2. **What is your desired future state?** Questions about innovation, serving new emerging markets, a profile of desired customers, launching new products or services, investments in new technologies help to clarify the new vision, and are you chasing or leading the competition on the way to the new vision.
3. **Define the gap between A and B.** What will be sacrificed, what will be gained, what will the acquisition of the new vision yield in the long run, what will it cost, is it worth all the effort?
4. **What is the optimal action plan to travel the gap from A to B?** Define the enablers and inhibitors that will get you from A to B. What are the steps of the plan, what is the critical path of the project, what are the contingencies if the plan goes off the rails?
5. **Evaluating the success of the plan**. This is an assessment of the metrics defined in the first step. Only the desired outcomes defined in the first assessment stage are valid measurements. To add, edit, or delete any of the criteria for success defined at the very beginning will undermine the efficacy of the strategic plan. If new lessons are learned along the way, then they can be included in later iterations of the strategic action plan.

Which brings us back to creativity and innovation. Professor Mark Allen observed in a recent presentation (2/18/2020) that creativity and innovation are separate functions. Creativity is the output of ingenuity and inventiveness. Innovation is the *implementation* of that inspiration. Some firms, like 3 M, may encourage creativity and take new ideas through the development process to the marketplace, thus encouraging employees to engage with the core mission of the firm and become partners in the commercial process in an open climate of participation and recognition. Other firms may be less open to creative ideas from employees. These latter firms are often less likely to sustain themselves in a competitive market as their customer base shifts steadily to newer innovations offered by other more agile competitive providers.

Those companies and countries who are open to creative innovation and who embrace a deliberate and rigorous strategic planning process will have a greater chance of success and long-term economic sustainability when adopting their commercial enterprise to a space-based operating model. It is impossible to anticipate and plan for every act of God and competitive tactic but those firms who have incorporated agility as a core strategic value have a better chance of accepting creative solutions and implementing innovative initiatives.

Preview of the Five Phases of Adoption of the New Space Economy

There may not yet be an official master plan for constructing the new space economy but there is a highly predictable and well-researched path to build the emerging space economy based upon Dr. Everett Rogers' *The Diffusion of Innovation* model (1962, 2003)) which will be more fully discussed in a later section. Based upon Rogers, the five predictable phases of adoption describe the path from an unorganized frontier economy to a more robust emerging economy and then finally to a strong and fully developed independent economy. As a preview, the five phases of adoption are:

1. Frontier—Innovators, risk-takers
2. Early Adopters—Infrastructure, volatility
3. Early Mainstream—Expansion and growth
4. Late Mainstream—Emerging markets and economic complexity
5. Lagging Adopters—Mature and autonomous markets in space

Later chapters in Sect. 5 will more fully present the five phases of economic adoption, discuss the likely business models that best fit each phase, and suggest the likely industries along with examples of individual companies that will thrive in the new space environment. I will show how these three elements can be integrated to create a clearer, more manageable, picture of how the space-based economy will develop and how entrepreneurs, investors, future space settlers, and corporate planners can evaluate when and how to enter the new space economy to maximize their investment returns and increase their chances for personal success.

As more companies and independent investors seek participation in the space economy the economic platform will become more robust, will stabilize, and will gain greater economic self-sustainability. The benchmark end-state will likely be when all economic activities, from raw materials acquisition, to production, and finally to consumption, will be transacted in a closed loop that is independent of Earth. In other words, the extraterrestrial market will be fully developed and mature when space-based producers will sell to customers in space.

Business Models in the Cosmos Economy ...

Successful business ventures don't just happen organically. To build and maintain a long-term sustainable competitive advantage the founders of an enterprise start with a draft business model that aligns to the core purpose and vision of the business.

Matching an appropriate business model with the right market is essential to mitigate risk for early investors. For example, Microsoft's early strategic choice to focus on software at the beginning of the PC boom while others focused on hardware, clearly differentiated their marketing terrain and allowed them to define an emerging customer base. IBM's decision to stall development and production of small personal computers (PCs) because they believed the market potential was too small led to their ultimate exit from the consumer computer marketplace. The innovative business approach of firms like Airbnb, Uber, Lyft, and others have redefined previously existing business models and transformed them to fit more closely with what customers really want—in this case the convenience of increased control of the experience, customized transactions, and a flexible price that matches the value proposition. But market disruption comes at a cost. For every Uber there are dozens of failed attempts and short-lived product launches that missed the mark of matching the moving target of consumer preferences.

A summary of effective business models (Wirtz et al. 2016) characterizes the formal business model as, "… a simplified and aggregated representation of the relevant activities of a company." Even more significantly the authors identified the key components of a successful strategically focused business model as having:

(a) The ready access to and use of resources including various levels of talent
(b) A clear understanding of the customer profile and the reasons of their demand
(c) A concise and executable value proposition that sustains long-term competitive advantage
(d) A strategic plan that accounts for the inevitable potential pivots that occur along any realistic growth curve
(e) A focus on revenue that feeds and underpins the long-term success of the enterprise

The biggest challenge when choosing the best business model to launch a business in space is defining the potential market. Who are the potential customers and suppliers in space? Is there adequate access to production resources such as raw materials, energy, or transportation? How will profitability, or the break/even point, be calculated in an undeveloped and evolving space economy? Will the lack of clarity about the details of the new emerging market inhibit investors and restrain strategic flexibility as things inevitably change while the market matures? And that's just for starters!

Determining when to enter the space economy at a specific phase with an appropriate business model is the result of a healthy mix of hard data analysis and intuition. Nobel Prize laureate Daniel Kahneman in *Thinking, Fast and Slow* (2015), suggests that effective decisions are a combination of logical and emotional components. (Left-brain *and* right-brain thinking.) While it is tempting to over analyze a proposition, Kahneman suggests the best approach is to include a healthy dose of intuition (thinking fast) to balance logical deliberation (thinking slow).

Of course, it will probably boil down to answering a few critical questions such as:

- Can the firm take a long-range view and support a long-term investment horizon or are investors adamantly wedded to a short-term ROI?
- Will the current set of capabilities (leadership talent, technology, culture, corporate secret sauce) mesh with this venture? This is another way of saying: How willing is the company to make, buy, or borrow new skills to launch and sustain this effort?
- With whom can the firm partner? Is the firm willing to forfeit some degree of ownership to mitigate the risks of launching the new venture in a new environment?
- Does the firm have a governance model that supports this venture? Will the board support commit to a long-term strategy?
- What's the most compatible business model to support this strategic decision?

Adapting to new markets is not a simple undertaking. Clayton Christensen noted in *The Innovator's Dilemma*, (1997, 2006) that organizations with established customers, supply chains, and market identities become wedded to doing business in a predictable way; it is their tried and true processes, after all, that got them to where they are and that defines their competitive position in the market. To make a radical change in their business process takes a conscious effort to transpose creativity into to implementable innovation. For many, the risk of leaving a loyal known customer base outweighs the potential gain of innovation.

Looking Upward, Looking Outward

The progression of the new space economy will follow a development path that will create a new suite of commercial opportunities in each successive phase of growth. In this dynamic setting I predict that there are some businesses that will have more potential success earlier (innovators and early adopters) and that others will benefit from a more stable and mature setting (late mainstream and lagging adopters) and thus find greater success in a later phase of the space economy. Given the huge potential of space, we can expect many frontiers evolving quickly or slowly into mature markets simultaneously in our solar system in the coming century(s). This means there will always be a frontier somewhere in space if there are risk-takers willing to boldly go.

Table 8.1 is a preview of the full 5-phase model of adoption. It combines the critical elements of optimal business model, and likely industries that will be a best fit for each adoption phase. What follows is a notional roadmap for entrepreneurs, investors, and enterprises to consider as they ponder the calculus of establishing a competitive presence in the new space economy. A fuller discussion of the five-phase model of adoption will be discussed later in much greater detail in section 5.

Preview of the 5-Phase Model of Adoption

The question of where profits are patriated will be another issue to resolve. Past colonial ventures were often able to keep profits away from prying tax collectors in their home countries by establishing local banking in the emerging market or via third party transactions. Whether or not this will occur in space is yet to be seen. The first phase of the space economy, the frontier phase, will afford opportunities for early entrants to define and build the foundational infrastructure that will help to enable later entrants and investors. Much of the economic activity of the early frontier phase will naturally focus on creating infrastructure platforms for communications, transportation, human habitat, food and water for those people who will have to live and work in space. I believe a milestone signaling the adoption of the frontier phase will be the establishment of a robust financial infrastructure.

Table 8.1 The new space economy will evolve in five overlapping phases of adoption. Each phase represents a successive era of economic activity that will benefit from prior ventures. Different business models, industries, and enterprises will align fittingly for each phase of adoption. Date ranges are notional

	Phase 1 Frontier (innovators)	Phase 2 Transition (early adopters)	Phase 3 Growth (early mainstream)	Phase 4 Emergent market (late mainstream)	Phase 5 Maturity (late adopters)
Approx. dates	2030–2060	2050–2075	2070–2095	2090–2115	2110 →
Description	No infrastructure, no consumer base, no industrial capacity, high investment potential, high entry/exit barriers	Significant economic activity causes tipping point from tentative nativity phase to solid growth phase	Broad economic base of investors and actors. Initial wave of consolidation results in more stable economic base, moderate entry barriers	Acceptance of sector as (a) normal, (b) integrated with earth economy, (c) viable, and (d) robust and profitable	Economic autonomy, M&A, space-based firms, P2P trade, very low entry/exit barriers, potential emergence of post-industrial economy
Likely business models	PPP, SOE, private capital	Open business model, cartel investors, VCs. PPP, SOE persist	Open business model, cartel investors, VCs, long tail model PPP fades	Traditional business models, long tail model, cartel investors, and VCs	Traditional business models
Likely industries	Communications services, infrastructure systems, robotics and AI, transport to/from earth	Human habitat and life services, energy production and distribution, engineering and construction, micro-gravity manufacturing, mineral extraction and processing	Food production and distribution, services (med, legal, security), shipping/depot/distribution, transport intra-space	Manufacturing—Traditional, tourism and recreation, transport—Planet surface	Consumer products/services, trade expansion, retail

References

Christensen, C. M. (1997, 2006). *The innovator's dilemma: the revolutionary book that will change the way you do business*. Collins Business Essentials.

Covey, S. R. (1989, 2005). *The 7 habits of highly effective people: a personal workbook*. New York, Simon & Schuster.

Kahneman, D. (2015). *Thinking, fast and slow*. New York: Farrar, Straus and Giroux.

Machiavelli, N. (2016). *The prince*. Value classic reprints. (Original publication circa 1500.)

Porter, M. E. (1980, 2004). *Competitive strategy: techniques for analyzing industries and competitors*. Free Press.

Rogers, E. M. (1962, 2003). *Diffusion of Innovations*. Free Press. (Originally published 1962).

Tzu, S. (2010). *The art of war/the ultimate guide to victory in battle, business, and life*. Best Success Books.

Von Clausewitz, Carl. (1997). *On war*. Wordsworth. (Original work published 1832).

Wirtz, B., Göttel, V., & Daiser, P. (2016). Business model innovation: Development, concept and future research directions. *Journal of Business Models, 4*(1), 1–28. Retrieved from http://www.businessmodels.eu/images/banners/Articles/Wirtz_Gottel_Daiser.

Part II

Coming into the Cosmos

This section considers the impact of enabling an unending series of new frontiers, describes the down to Earth problems of distributing goods made in space, reviews how past commercial enterprises took risks and built colonial empires in the underdeveloped economies of the world, considers reasons for resistance to the forward momentum of building industrial space, presents the dominant visions for the future of commercial space, and discusses how privatization of space may serve as a primer for overcoming challenges to the new space economy.

9

Settlement Communities

When Frederick Jackson Turner presented his theory about the closing of the western frontier signaled an end of American civilization's economic expansion in 1893, he would have been surprised to know that lessons from his *Frontier Thesis* have persisted well into the twenty-first century. Notably, the image of the frontier as a metaphor for human progress was invoked by John F. Kennedy in his presidential nomination acceptance speech at the Democratic party convention on July 15, 1960. At that time candidate Kennedy (1991) famously said:

> We stand today on the edge of a New Frontier—the frontier of the 1960s, the frontier of unknown opportunities and perils, the frontier of unfilled hopes and unfilled threats. … Beyond that frontier are uncharted areas of science and space, unsolved problems of peace and war, unconquered problems of ignorance and prejudice, unanswered questions of poverty and surplus.

The narrative about the importance of the frontier that started with Turner and made its way to Kennedy is not coincidental. When the speech was delivered, there were still people living who made the trek west by Conestoga wagon and who had memories of growing up on the Western frontier. Kennedy and Ted Sorensen, his speech writer, understood that the romantic imagery of the frontier was very much alive in the minds of the mid-twentieth century citizens and that evoking ideals and values embodied in the "New Frontier" signaled hope and optimism to American citizens as well as foreign nationals seeking a better life of opportunity. Much of President Kennedy's political legacy was embodied in his New Frontier programs aimed at social and economic revitalization. A key element of the New Frontier was a commitment to space exploration and technology resulting in the Apollo program that put American astronauts on the Moon in July 1969.

By going to the Moon, we challenged the idea that the frontier ethos had disappeared. Turner may have defined the importance of the frontier by eulogizing its passing, but Kennedy resurrected it, gave it a socio-political rebirth, and set the stage for manned exploration of space. His legacy is that the promise of the new frontier of space remains vital and real, thus opening the door to establishing future settlements throughout our solar system.

JFK's New Frontier agenda was broader than just space exploration and included a variety of legislative initiatives aimed at solving endemic social problems like civil rights, education reform, environment, health, and unemployment. Because space exploration was included on Kennedy's list of New Frontier priorities it was easy to assume that, like the social programs on the list, space and technology were there to solve social problems. The idea that space exploration has value, because it serves a fundamental social need, was amplified by its inclusion as an integral part of the New Frontier agenda. In so doing space officially graduated from science fiction to science reality and was recognized as a social benefit to humanity.

Consumerism and Consumption

Here in the U.S. and most of the world, consumerism is taken for granted as the natural way of life. But the acquisition of things as a core societal value is a relatively new idea. Before the industrial revolution, most people lived in poverty and were unable to purchase things that weren't essential to survival. If something wasn't directly related sustaining the basics of living it was a discretionary purchase; something frivolous (like a second pair of shoes or glass window panes). Peasants living in perpetual poverty only purchased the necessities for making food, clothing, and tools they were unable to make themselves. Only the very wealthy were able to purchase new things regularly without concern for their practical use. American consumerism, which has spread to much of the planet, holds that acquiring things—often things not critical for survival—is a core societal value. Going shopping is a recreational activity. Current fashion trends, not utility, often dictate what will be purchased. Acquiring more and more "things" and other extreme behaviors have become part of accepted social practice (like hoarding). As a result, a very successful business model is renting storage space to people who have acquired too much stuff to keep in their home or apartment due to their addiction to acquisition.

Things have become something more than merely practical objects. A pretty ornamental plate may become a wall decoration instead of a place for food, shelves may become cluttered with glass figurines, garages may house expensive recreational toys like snowmobiles, boats, motorcycles, and other icons of wealth, class position, or self-image.

Getting more and more stuff is decidedly *im*practical in space. Consumerism is not something that will translate to the environs of space settlement where people will live in close quarters with limited ability to possess items unless they have practical utility or serve the necessities of living and working in space. In this setting the private ownership of "stuff" will take on a different role and will probably be acquired much differently. Living in space will likely begin as a communal affair with ownership of durable equipment like shared transportation (instead of owning individual cars), the use of common resources for eating (communal kitchens), bathing (public baths), and other domestic activities. In this common setting the use and purchase of goods will be a cooperative transaction, a collective exchange where the purchasing decision will be made more for the good of the community and less for the pleasure of an individual.

An example of how commerce will persist without purchasing more stuff is the growing trend of a shared economy. Ride sharing with Uber or Lyft, taking a trip along with a stranger in their private car to get from one place to another, reduces the greater need for car purchases. House purchases are on the decline as demand for rentals has increased, using Airbnb to stay overnight in a stranger's spare room has become an acceptable way to travel, and renting entertainment on Netflix has replaced purchasing a tape or CD of the movie. Textbooks are regularly rented by college students on Chegg instead of purchased at the campus bookstore. At the end of the day the Airbnb host still owns her spare room and is able to rent it again, the college bookstore still owns the textbook and can rent it to another student, and the Lyft driver still owns her car and is able to keep on collecting fees for sharing rides all over town. Netflix doesn't have to keep hundreds of copies of old Jerry Lewis movies in inventory because just one stored on the server is quite enough! Oppenheim (2010) proposes that examples of disruptive "**Collaborative Consumption**" indicate a trend of non-traditional consumers who don't necessarily need to access goods or services in a traditional transactional way. Settlers in space don't need to bring photo albums of friends and relatives to space because they can always view the digital version. An extension of this model may find its way to bigger ticket items like furniture and household appliances which may be leased or rented along with the house or apartment, as is the case already in some upscale apartments in Tokyo and other urban centers.

Another alternative to consumerism is the American Indian tradition of potlatch. This approach encourages a person to become purposely unattached from their belongings and to give them away to friends and strangers as a sign of spiritual cleansing and social confidence. When an item becomes too important in someone's life, it can take on a strength of its own. This reverse view of the values we give to the things we own may liberate owners from the tyranny of their possessions. There is a Zen maxim that says, "The more possessions you own, the more your possessions own you." This is true in the world of ascetic monks and in the challenging future world of frontier space settlement.

Alternate Settlement Systems

Space settlement often evokes fantasies of an ideal future where all the problems on Earth are engineered away, where problems caused by conflict, rivalry, scarce resources, and all the other causes of stress and discomfort have been magically removed. Many of these illusions come from the disappointing reality of living here on Earth where status and wealth disparity produce societies of inequity; haves and have-nots. But the promise of space settlement is not the promise of a promised land. It is impossible to predict how various space communities, colonies, settlements, or outposts will be governed. Some will conform to their sponsoring nation or corporate social culture. Others will carve out their own unique social system of values and behaviors that best suit their shared objectives. In addition to social systems that may be purely socialist or capitalist the example below offers a hybrid approach, named community capitalism, that blends rewarding individual ambition with balancing community norms. In time this and other social forms may proliferate in space.

Community Capitalism is an outgrowth of China's efforts to democratize the ridged soviet style communist approach to centrally managing local economies. Hou Xiaoshuo, a professor at St. Lawrence University, has documented this three-tiered system of collective shareholding of for-profit enterprises.

This mandatory system (no one can opt out) of community prosperity is equally distributed among participating citizens like shareholders in a U.S. company who receives periodic dividend payments. Dr. Hou Xiaoshuo (2014) indicates that in this system, there are three ways a participating citizen can receive a monetary distribution. The first form of income is via a traditional "communist" *stipend* that covers foundational subsistence and day-to-day expenses like food. This is much like a guaranteed universal basic income (UBC) stipend that is rapidly gaining political and social acceptance in Europe and other areas and is designed to counter lost or reduced income due to job automation or economic recession. The second form of income is the "socialist" element where an individual receives a *salary* for work at a job. The last form of income is the "capitalist" component where the worker receives proportional income from shares and other investments in community ventures.

The hybrid system has gained wide acceptance in China because it allows for individual choice, can be altered to meet local community needs, and incorporates a multiple variable approach (communist, socialist, and capitalist) that can be altered and adjusted as economic situations fluctuate and change over time. A unique social benefit is that the mandatory nature of the social contract promotes a strong sense of community solidarity, supports individual achievement, and builds a work ethic ethos that becomes a stabilizing value in the community. A multi-system approach to economic sustainability is designed to benefit both community and individual in a mutually beneficial partnership—a social contract. As space settlements and outposts look to define themselves and define how members of the community will work and live together this model is likely to receive further examination. Because the origin of this model is Chinese, it will surely be employed in Chinese space colonies and may likely migrate to other non-Chinese space outposts and communities if it proves to be equitable and effective.

Currently, there is much conjecture about the rise of a post-capitalist economy due to the expected impact of automation on jobs at all levels of the social structure. The assumption is that artificial intelligence (AI) and automation (robots) will soon replace human labor for individual jobs and entire employment categories across the economy resulting in mass disemployment that will force a restructuring of the global economy. There is a range of proposed models describing the future of world capitalism in a highly automated setting where robotic production reduces scarcity, profit margins all but vanish, and the middle class is eviscerated. None of these views are optimistic. All these dire future visions base their assumptions on nineteenth century economic assumptions that supported Marxist theory which encouraged a hive-like approach to life while avoiding the human desire for individual achievement and recognition. One such example is *The Resource-Based Economy and the Collaborative Commons* (Dew 2020).

Resource-Based Economy and the Collaborative Commons. This exceedingly extreme post-capitalist utopian view assumes that all human labor will be replaced by automation resulting in an economy not based upon specific corporate profitability but on an egalitarian distribution of goods and services with little concern for merit-based contributions or social responsibility. The model assumes that once technology replaces all need for human labor, the current form of capitalism, based upon consumption and profitability, will be "washed away." Strongly dependent upon Marxist theory, this schema ignores the human nature of the individual and how personal ambition and response to motivating incentives and rewards would tend to pit the individual in opposition to the dictates of the social system and thus undermine the idealistic and simplistic all-for-one-and-one-for-all society.

Communities and Communes

For an older generation weaned on the promise that all things are possible (as long as there is the *right* leader who has the *right* vision), there is hope that space will provide a utopian alternative to life on Earth. In the past, during the American western migration of the nineteenth century, the belief and aspiration to find a more perfect life motivated communities of hopeful people to remove themselves from the greater society in order to live a more ascetic life in a cloistered settlement apart from the fast paced, economic centered, modern world of questionable values. Many of these social experiments were often characterized as religious cults or offshoots of mainstream beliefs. Just like the social experiments of the nineteenth century in the American frontier, some experimental communities were social utopians and others chose the cloistered life as a rejection of civilization and the imposition of modern values (i.e., being acquisitive and industrial). Examples include the Amish of Pennsylvania who are still a vibrant community and the Shakers founded in 1783 whose current population is 2.

There were also counter-culture communities formed to model a more perfect ideal of what society was *supposed* to be. In the 1960s, there were numerous communes that sprung up where ownership of goods was disdained, monogamous relationships were rejected, and work shared. An example is The Diggers who established their commune in San Francisco's Haight Ashbury district from 1967 to 1969. The original setting of the long-running comic strip, Doonesbury, was about characters who were members of a commune called Walden patterned loosely on The Diggers.

Not every utopian community espoused the better values of society or the spiritual side of human nature. Miles Harvey in *The King of Confidence* (2020) chronicles the Strangites, a breakaway offshoot of an earlier utopian society of the mid-nineteenth century led by James Jesse Strang, a charismatic character who declared himself king of their isolated upper Midwest enclave of Beaver Island. Harvey describes the group as a "theocratic kleptocracy" due to their reputation for engaging in piracy, counterfeiting, and other assorted unneighborly crimes. At one time the group had as many as 2500 reported adherents. Today there are still may be about 130 loyal devotees.

A major theme of Frederick Jackson Turner's thesis about the impact of the closing of the American frontier was that once the frontier wilderness was gone, opportunities for outliers in society, those who didn't fit with mainstream thinking or behaviors, had no place to remove themselves except to live in urban fringe areas. Without the safety valve offered by the frontier back country, these nonconformists, eccentrics, and contrarians would have to navigate within an increasingly confining and restrictive civilized world. A wonderful example of the conflict caused by this struggle of conflicting ideas is the movie, *Lonely Are the Brave* (1962), written by Dalton Trumbo and staring Kirk Douglas. The hero is a Korean War vet (Douglas) who rejects society by living job to job, day to day out on the fringe of society. In the film he temporarily leaves his itinerant life as a cowboy on the range (i.e., the frontier) to try and live in a restrictive rule-heavy civilized world. The film shows how society tries to crush his spirit, his individuality. Douglas long maintained that this was his favorite film.

The prospect of life in space offers a chance to build new communities apart from the rest of civilization. As reported by McClure in *Cosmopolitan* (2020), things can go terribly wrong even in utopia. The story reports over 152 cases of sex abuse and incest were documented by the Amish community of Pennsylvania. Human misbehavior can't be magically improved just by isolating the group from the rest of humanity. Space utopians take note that a life on a settlement millions of miles from Earth is not a guarantee of a better life no matter how far removed you may be from the rest of humanity's perceived ailments.

References

Dew, C. (2020, April 27). *Post-capitalism: Rise of the collaborative commons*. Medium. Retrieved from https://medium.com/basic-income/post-capitalism-rise-of-the-collaborative-commons-62b0160a7048

Harvey, M. (2020). *The king of confidence: A tale of utopian dreamers, frontier schemers, true believers, false prophets, and the murder of an American monarch*. Boston: Little Brown and Company.

Hou, X. (2014). *Community capitalism in China: The state, the market, and collectivism*. Cambridge: Cambridge University Press.

Hou, X. (n.d.). *The curse revisited*. Retrieved June 15, 2020, from http://assets.cambridge.org/97811070/30466/excerpt/9781107030466_excerpt.pdf

Kennedy, J. F. (1991). *Let the word go forth: the speeches, statements, and writings of John F. Kennedy 1947–1963*. Laurel. (p. 101).

Lonely Are the Brave Full Movie. (1962). YouTube. Retrieved June 19, 2020, from https://www.youtube.com/playlist?list=PLcS8cosD056sT5_QWpcMPy9SlirN5bcGj

McClure, S. (2020, January 14). The Amish keep to themselves. *And they're hiding a horrifying secret*. Cosmopolitan; Cosmopolitan. Retrieved from https://www.cosmopolitan.com/lifestyle/a30284631/amish-sexual-abuse-incest-me-too/

Oppenheim, L. (2010, September 21). What's mine is yours—the rise of collaborative consumption (book review). TreeHugger. Retrieved from https://www.treehugger.com/culture/whats-mine-is-yours-the-rise-of-collaborative-consumption-book-review.html

10

Forever Frontiers

Settling the solar system will take a very *very* long time simply because our solar system is a very *very* big place. Generations from now, new frontier settlements will still be starting somewhere in the solar system, an endless nursery for new communities. Some of those communities will be affiliated with other like-minded settlements because of shared political or religious beliefs. Others will connect through trade and commerce. There might even be those space shtetls whose culture and mythos may be traced to some common ancestry back on Earth. Others will emerge fully independent of Earthly traditions.

Continual, sustained adoption of space enterprise will be important for several reasons. First, it is central to the notion of expanding the space-based economy. Just like continually opening new stores in a commercial chain (e.g., Starbucks, Taco Bell) promotes brand awareness and penetration to more customers, the continuous expansion of the cosmos economy increases the diffusion of business providers, and consumers.

Second, the prospect of a practically endless supply of new frontiers, and the new enterprises that seek their commercial success on these frontiers, means that the industrialization of the new space settlement economy will have steady footing for generations to come. This is good news for entrepreneurs and investors who traditionally seek a first-mover advantage when establishing an enterprise in a new or untapped market.

The third is if an investor missed the computer revolution, or the digital revolution, or if they failed to see the opportunities of the internet and missed the boat because the boat had already sailed by the time they finally figured out what everyone else was so excited about, then knowing there will be a nearly endless supply of frontiers, with new opportunities, and new markets to stick a cautious toe in, is very good news. As we shall see in later chapters, new markets occur in progressive phases of adoption, represented by levels of risk and expected return. The first phase of market adoption, the Frontier Phase, is where most risk resides and the first frontiers to develop in space are the riskiest. Naturally, the cautious risk-averse investor is typically reluctant to engage at this early level. But if there is a steady supply of new frontier markets on the horizon, then the normally cautious entrepreneur may be more likely to engage in a new frontier economy. The result of mitigating risk at the early stages of market adoption is a stronger, integrated space economy overall.

Some businesspeople are naturally cautious about investing in new things. The very idea of investing in space could easily move someone far beyond their comfort zone. Especially if the highest risk for an investment is in an untried frontier out in space. But that very scary investment risk could become far less frightening if it could be demonstrated that others have gone down the same path and profited. For risk-averse investors who are more comfortable making an investment in something safe and well-proven—like real estate—then the idea of another frontier isn't so worrisome. Especially when the investor considers that frontier markets can quickly grow into mature markets thus giving the frontier investor the benefit of being an early mover into a new settlement market.

Waiting for Anti-gravity: Doing Down-Mass

Transportation to and from space is imperative for the adoption of the overall space economy. But this is easier said than done. Space-based commercial activities like micro-gravity manufacturing, of pharmaceutical production, heavy machinery manufacturing, hazardous industrial production off-planet, farming food and fiber for the planet's increasing population, and mining and refining raw resources from space all need clear distribution to market. Without capable distribution channels there can be no trade, and without trade the space economy may falter or fail. For now, Earth is where the customers are—there are no consumers yet in space—it stands to reason that all those goods produced in space must be delivered to customers down to Earth. But there's a glitch. Today there is no capability for bringing all those tons of goods from space manufacturers to Earth's markets. Astronauts, space tourists, and small payloads may be occasionally returned to Earth, but these are very light loads. We are limited to shipping only the rough equivalent of a small truckload—not enough cargo to make a dent in the commercial market. If space is to evolve into an economic powerhouse, the ability to transport thousands of tons of cargo—not just a few hundred pounds—down to Earthside spaceports is critical. Unfortunately, this seemingly simple capability has not yet been developed and is unlikely to be resolved for decades to come. Dylan Taylor, CEO of Voyager Space Holdings, a conglomerate of complimentary space companies, refers to this as the down-mass problem. Whenever I brought up the down-mass problem in an interview or at a space conference I always got the same reply: "Someone is surely working on this problem, but I don't know who that is."

By comparison, the ability to ship commercial space cargo at a similar volume to that of a major U.S. port does not yet exist. For example, the Port of Long Beach, California, manages an estimated 193.1 million tons of cargo annually (2019 data). And the Port of Long Beach is just one of dozens of similar ports around the world. If the space-based economy is expected to generate enough revenue to attract investors and compete with Earth-based enterprises, then goods will have to flow with similar ease and volume from space to Earth.

There are no readily available realistic remedies to this problem. There are, however, several very *un*realistic proposed solutions that keep resurfacing in conversations with people who are considered experts. I will include three of them here. Disclaimer: Normally I would discount the following three solutions as pure fiction, but I am including them just in case one of these plans becomes reality and I am proven wrong.

The first example comes from Star Trek: teleportation. To be fair, Star Trek didn't invent this future technology; many of the ACE paperback Sci-Fi books of the 1950s used some form of *beaming* things and people across great distances. If this capability were to be developed, then Amazon Prime members wouldn't have to wait a whole day for their deliveries but could get everything immediately. There is certainly a tremendous commercial incentive for beaming things from place to place. Likewise, beaming people around the planet seems like a logical next step. Right now, the best replacements for air travel are teleconferencing with Zoom™, email, and simply refusing to fly on a plane with a bunch of rude strangers. [The coronavirus pandemic of 2020 heightened passenger fears about the toxicity of all the sneezing, coughing, scratching, and other activities going on in the next seat.] If beaming works for inanimate things like paperclips or tugboats then it will only take a few minor adjustments to make it safe for people, pets, produce, or plants to make the trip. Stay tuned.

The next super technology not yet developed but ardently hoped for by science buffs is antigravity. The ability to harness one of the prime forces of the universe for common commercial purposes would certainly be a crowning achievement for our civilization. In addition to the much-appreciated cultural benefit of putting Las Vegas magicians out of business [I know how you levitated that person: antigravity!], the prospect of navigating the new normal of a world that no longer recognizes concepts of neither up nor down would leave many of us perpetually up in the air.

Finally, there are other, less fanciful, schemes for connecting Earth to space. At the top of this list is the long-anticipated development of a space elevator. As the name suggests, a space elevator is a physical connection between a geosynchronous platform up in space and terminal location down on Earth's surface. The manner of the connection, either one long super-strong cable (or filament) upon which goods and people could be conveyed or a balanced pair of connections allowing one conveyance to descend while its mate ascends, has yet to ascend from the drawing board to reality. If this technology were successfully developed, then terminals at spaceports all over our planet would become gateways to space and space goods would find distribution to markets all over Earth. A space elevator terminus near most of Earth's major cities would mean that goods could be transshipped all over the planet with routing logistics occurring in orbit. This would replace most long-distance rail, truck, and ocean cargo (and passenger) activities. Unfortunately, space elevators are still suffering the ups and downs of developing a working proof of concept prototype. Otis take note.

But while the down-mass problem currently inhibits direct trade from space to Earth, it will ultimately enable another set of lucrative economic opportunities. Stay tuned.

A Fork in the Road

If the new space-based economy is to grow and foster a robust platform for continued commercial space adoption, then there must be a viable financial reward for investors seeking to support and to profit from this new arena. It boils down to who are the customers, consumers, and corporate clients, what are the goods that will meet their demand, can those products be efficiently produced in space, can those goods be delivered economically, and can this value chain be expanded (i.e., scaled) to accommodate anticipated increases in demand so that producers can benefit from economies of scale and other production efficiencies? All at the right price point.

A Fork in the Road to Space

Plan Alpha. If antigravity, teleportation, space elevators, or some other technologic leap of logic can be successfully implemented, then customers on Earth will eagerly respond to space-sourced products. The type of goods manufactured in space will be defined by customer demand located on Earth. Just as manufacturers in China seeking American markets readily produced Christmas lights and ornaments even though there was zero demand for these items in China, space-based manufacturers will tailor their inventory to fulfill customer demand on Earth. Business between space providers and Earth customers will likely be heavily transactional and less reliant on building long-term relationships. With open trade access to Earth's global markets the space economy could rapidly grow at an accelerated pace and fuel further investments in space-based enterprises.

Plan Beta. But if none of the Plan Alpha technical solutions come online (or until something does), then the development of the space economy will take a different direction and a different pace of growth. In this scenario, customers in space, instead of on Earth, will call the shots. Space customers will have a different demand profile: less of a consumer base and more of an industrial definition of demand, so the output of space-based manufacturing will conform to a more commercial set of customers located in space and become more oriented to creating products that serve utility and practical application (especially for industrial customers), for activities like building infrastructure. With limited trade with Earth the early business-to-business (B2B) space economy will grow at a much slower pace, a more deliberate rate.

Either scenario, plan Alpha or Beta, will enable the development of a new space economy. Plan Alpha will allow space to develop on an accelerated trajectory while plan Beta will be more deliberate but will also be more likely.

Solar System 101

It's easy to forget how vast our solar system is. Maybe this is because we are taught about the planets spinning around the Sun using a simple mechanical model that suggests the whole thing can fit on a school desk or that the planets are all in convenient reach of one another. Maybe it's hard to grasp the size of the solar system simply because it is bigger than the limit of our imagination (Redd 2017). For most of us there is no conceptual difference between a billion and a trillion; they are both ginormous numbers that are too big to fathom with no practical relation to everyday life. Maybe that's why we underestimate the full potential of tens of billions of people living in hundreds of thousands of settlements and outposts in our solar system in the millennia to come.

Our continuously expanding universe is estimated to be about 28 billion light-years across but since the exact shape is not known, neither is the exact measurement. Because each light-year is 5.879 trillion miles, the universe is 164.612 billion trillion miles across. That's an enormous size to consider. If the idea of visiting other planets and celestial bodies like asteroids and planetary moons in our solar system isn't headache enough, add to the mix that NASA estimates over four thousand exoplanets have been identified as of this writing (Mack 2019). Someday, if we can overcome the physical limit of not traveling faster than the speed of light, our decedents will likely pay them all a visit. In the meantime, we have a whole unexplored solar system chock full of all sorts of places to settle.

Our solar system, just one of several million similar small specs twirling endlessly in our galaxy, is estimated to be 7,440,000,000 miles across. If these very large numbers describing the size of the universe and our solar system don't make real sense to you, don't worry. The incomprehensibility of these immense cosmic distances is just a limitation of our tiny human brain that is accustomed to thinking in terms of the distance to the grocer or how long it takes to drive to visit friends. Longer trips are thought of in terms of time; the estimated hours or days it will take to drive to the lake house for the weekend or how long it will take to fly to Tokyo (or from Tokyo to LA). Just a few short centuries ago most people spent their entire lifetime within a short day's walk of their home village, and it took months to cross the Atlantic. In a few short centuries from now people may look back at us and marvel that people living on Earth spent their entire lifetime on the same small planet. For now, only a few people have learned to think in terms of cosmic distances. Maybe, in time, you will think this way, too.

You Don't Miss your Water 'Till the Well Runs Dry

Water is critical for human survival. Not just for hydration, but also for growing crops in space, making fuel, among other uses. Beyond our solar system, over 4000 exoplanets have been discovered so far, and at least a third of them are made up of at least half their mass as water. Here in our local solar system, water is more abundant than originally believed just a few short decades ago. While Earth is the only major planet in our solar system with surface liquid water, there are other plentiful sources for water—most of it in the form of ice

(Lovett 2018). Sources of liquid water can be found on moons Ganymede (46% liquid water), Titan (26%), Callisto (9%), Europa (16%), and even on Pluto (15%). By comparison Earth's surface is 72% liquid water but is much larger than these other bodies. Ceres, the largest asteroid in the solar system, is believed to also have oceans of liquid water. But this has yet to be confirmed. There is water in the form of ice at the polar regions of Mars and the Moon. The Artemis program plans to locate its base at the southern Lunar pole to access the deposits of frozen water.

In addition to the well-known eight major planets (sorry, Pluto) and all their many moons, recent observations have located 139 new so-called minor planets adding to the 245 bodies already cataloged (Tangermann 2020). These "tweener planets"—too small to be classified as a major planet, but too big to be either a comet or a space rock—have been tracked in orbits beyond Neptune in the Kuiper belt region. The solar system is so large that we will probably continue to discover new minor planets and other bodies for a long time to come. Each of these holds the promise of new knowledge, and new opportunities for settlement and commercial enterprise.

Given the size and complexity of the solar system the down-mass problem may not be a problem after all. If the first several decades of the space economy are spent developing and serving industrial customers and consumers in space—instead of catering to Earth markets—it is likely that the space economy will flourish and grow in pursuit of the demand that comes from increased exploration and settlement in the solar system frontiers and take on a character all its own.

References

Lovett, R. A. (2018, August 28). *Water worlds are abundant in the universe, researchers say.* Cosmos Magazine. Retrieved from https://cosmosmagazine.com/space/water-worlds-are-abundant-in-the-universe-researchers-say

Mack, E. (2019, July 10). *NASA drops insane map of 4,000 planets outside our solar system.* CNET. Retrieved from https://www.cnet.com/news/nasa-drops-insane-map-of-4000-planets-discovered-outside-our-solar-system/

Port of Long Beach. (2019). *Tonnage statistics.* Retrieved from www.portoflosangeles.org, https://www.portoflosangeles.org/business/statistics/tonnage-statistics

Redd, N. T. (2017, June 7). *How big is the universe?* Space.com. Retrieved from https://www.space.com/24073-how-big-is-the-universe.html

Tangermann, V. (2020, March 16). *Scientists discover 139 new minor planets in our solar system.* Futurism. Retrieved from https://futurism.com/scientists-discover-139-new-minor-planets-solar-system

11

Pushback and Challenges

Not everyone is in favor of creating a new settled commercial space economy. There are those who actively resist this development for economic, social, or political reasons. Competitive industries will resist because space will offer a new source of competition for their products. Political opposition will likely come because new outposts and settlements in space may challenge the sociopolitical status quo. Those with social rank and privilege will resist the development of space because their social standing may be devalued as new communities and social experiments emerge on the space frontier. Skittish investors may balk at funding space start-ups because of the high risk and uncertain return.

Balancing arguments against space settlement is the central purpose that drives commercial enterprise: a personal vision of success, often in pursuit of profit and power. Embarking on a path to settle space must surely acknowledge a list of potential antagonists who will seek to undermine prospects for success and evoke strategies and supporting tactics to thwart start-up efforts. Here are just a few probable challenges that space entrepreneurs will need to address in their business plans.

The challenge of the start-up. Early adopters of the space settlement economy, like early adopters in other sectors, will face a variety of tests that later adopters may not have. These entry barriers to business in space will increase the cost of the enterprise and augment the risk of failure. Capital used to overcome "external" concerns like the lack of infrastructure, the challenges of inadequate supply chains, competitive rivals, conflicting jurisdictional regulations, and other challenges may come at a high price. The opportunity costs associated with spending time and money on non-core activities may be better deployed on "internal" operational development instead.

But, a steep barrier to entry may be an asset. Once an enterprise has crossed the start-up threshold, it can claim the territory, the new market space, as its own. Others who follow will surely draft in the wake of first movers. They will have the advantage of learning from first-mover mistakes, but they will have to compete as a market follower. The early adopter, the market leader position, is an advantage that can be leveraged with tactics like premium pricing and plans for expansion to other space markets, for example. Being the market leader provides the early mover with unique advantages due to the *Principle of Cumulative Advantage* which suggests that the early mover will gain compounded benefit over time, ahead of competitors who enter the market later on, just because of their ability to capture market share with little constraint. Market leaders also have the advantages of brand recognition, customer loyalty, leveraging production economies of scale, and price limiting which drastically reduces profit margins and may serve to constrain competitor entrants who cannot match the market leader's production efficiencies. [Note: limit pricing is not the same as "dumping" which is the practice of selling goods below the sum of all production costs.] Sometimes the challenge of the entry barrier is only half the story. The firm must pay attention to the cost of exiting a market as well. Sunk costs in equipment, licensing, physical plant, and other expenses may prohibit a swift and painless withdrawal from an untenable commercial venture.

The challenge of competitive rivalry. U.S. firms will not be the sole or even the dominant actors in space. Other companies, countries, and consortia will play very active roles in promoting their own version(s) of a space economy. The rivalry between competing nation states and private investor cartels may be healthy or may cause the groundwork for political and military conflict between settlements of competing sponsorship. With the promise of high reward for high-risk investments in space it would be no surprise to see competing factions create social, political, economic, technologic, and legal barriers to protect their interests and to maneuver independently without the constraint of oversight and criticism from rivals.

The challenge of technology. In the early phases of adoption (discussed in greater detail in Chap. X), the underlying technologies needed to establish settlements, create a viable commercial economy, and promote trade, may not yet be fully acclimated to space. Technology developed from an Earth-centric perspective may not adequately serve the needs of space-based industrial operations. This gap of Earth vs. space technology will close in time as the scope of demand in space becomes clearer. An example of such a technology gap is the current lack of ability to ship large payloads of manufactured goods and other materials from space down to Earth. As of this writing, there is no developed capability to bring significant tonnage to consumers and industrial customers on Earth because of this down-mass problem. The consequence of this technological gap is that producers in space will have to look to consumers and customers who are also in space instead of the more lucrative mass consumer base on Earth. Products targeting space-based consumption will be fundamentally different from products designed for sale through Earth's distribution channels.

On the other hand, if this technologic gap is closed, then products produced in space will be designed to serve different customer criteria in a variety of Earth's market sectors.

An additional technology challenge will likely be creating compatible standards across various rival national and corporate platforms. Whether or not intellectual property (IP) is shared or hoarded will be the basis for either future cooperation or competitive rivalry.

The challenge of limited infrastructure. Just as there is no consumer base in space—thus hampering those investors with hopes of profiting from traditional business models—there is also no infrastructure necessary to support new business. This lack of utility is both a curse and a blessing. The commercial and social needs of businesses, communities, and individuals in space will have to be supported by a platform of shared infrastructure. The advantage of starting from scratch is that investments in building infrastructure will be tailored to the beneficiaries as the economy grows and develops. This will create opportunities for entrepreneurs to provide services in areas like transportation (both in space and on the surface of a moon or a planet), construction, insurance, banking, energy production, habitats including individual housing as well as manufacturing and space-based industrial facilities, warehousing and logistics, life support including food and water, medical services, and general store mercantile retail channels.

The challenge of frontier justice. Unlike our current western notion of individual rights supported by a fair and just legal system, the space frontier will likely lack fairness and justice for some time to come. In this regard the space frontier may resemble the American western frontier of the nineteenth century. Space, like the Old West, will likely be a wide open place devoid of ready accountability for crimes committed. The vast distances between settlement communities and far-off outposts will challenge local inhabitants' ability to mete out justice for personal and industrial crimes. In this arena the "might" of the company may easily squash the "rights" of the individual.

To further complicate the enforcement of legal fairness, the different rules and laws of one corporate outpost may not be recognized or apply at another national outpost where there may be neither reciprocity nor extradition agreements. In the plot of the film, "Butch Cassidy And the Sundance Kid," Robert Redford and Paul Newman were able to evade American justice by first seeking sanctuary in Mexico and later escaping to Bolivia. Escaping across the border is a well-worn trope of fiction. While we're on the subject, science fiction is ripe with examples of space crooks, pirates, and schemers who use this ploy. Borders and territorial demarcations of control may be hard to map in the rapidly changing setting of industrial space; it will be relatively easy to inadvertently cross these lines only to discover that assumptions about what is or isn't legal may no longer apply.

The challenge of Robots vs. People. Space today is truly no country for Earth men. There is no doubt that the early stages of space settlement will be dominated by AI and robotic technologies. A template for this is the Chinese method of establishing a scientific station on the far side of the Moon. The first phase of construction was completed via automation, without the help or intervention of humans. In time, the goal is to create a facility (they are politically careful not to refer to this construction as a "base" or "outpost") that will accommodate human visitors and, it is assumed permanent staff. This approach employed by Chang'e-4 makes sense. Robots are expendable, people are not. Robots can work continuously if their batteries are recharged; people need food, rest, medical support, and periodic bathroom breaks. Robots can be easily reprogrammed to accomplish new tasks when an old project is complete, but retraining people is not always as easy or as effective. More importantly, when people play an integral role in a hazardous setting like space the primary focus of the project shifts from the goals of the project to the safety of the personnel. When robots are involved the primary focus of the project is only the accomplishment (and cost) of the project, not the safety of the robot.

The use of automation and robotics to do the heavy lifting of construction and other jobs usually relegated to human labor further defines the types of roles that humans will play in space. There will be little need for human ditch diggers, mechanics, or other manual laborers in space. These and other tasks will be virtually eliminated by prefabricated habitats and assembled by specialized robotic workers built and programmed for this purpose. Humans will play only a minor role. People who will find a role in space will offer different talents: more technical, more strategic, more integrative, more intelligent, especially during the early phases of settlement adoption.

The challenge of monopolistic competition. If your vision of the future space economy is an open free market, then think again. Those enterprises that challenge and overcome the steep entry barriers to establish a commercial operation in space, either privately funded or sponsored by a sovereign state, will probably hold the high ground for some time to come. Their early-mover advantage will allow them to build an economic moat around their supply chain (or create a proprietary vertically integrated supply model) and then charge high-margin premium pricing for their products. In this way mini monopolies that differentiate on quality or other attributes that inhibit substitution with a competitor will likely dominate the early phases of the space settlement economy. These enterprises may employ such business models as subscription and leasing to further capture customers to their orbit as a defense against rapidly following competitors on the horizon.

As issues of limited infrastructure diminish over time, and as more and more businesses establish themselves in the market, it is likely that the first-mover advantages will likewise diminish. Later entrants, "fast-followers," and competitors will probably benefit from the lessons learned by the early movers as they find their own innovative path to success. If they are successful finding a winning formula, then the hegemony of the first movers will be threatened unless they are able to redefine their business proposition in a way that aligns to the needs of a rapidly changing customer base. At some point the likelihood of consolidation and M&A activities will probably topple the mini monopolies and create an economy more open to competitive pricing and free trade (Monopolistic Competition 2019).

Just as trading posts and general stores were the first seedlings of settlement in the American wilderness, often providing essential goods at premium (monopolistic) prices, so, too, will outposts be established on the ever-expanding fringe of the solar system's industrial frontier. It will take some time for competitive open markets to emerge in central trading hubs like the Moon, Mars, and other solar system commercial crossroads. Because the solar system is big, very big, the successful establishment of a frontier market in one sector, along with the mini monopolistic advantage of first movers, will probably propagate to other frontier markets in dozens, if not hundreds, of remote locations in space.

The challenge of space itself. The real cost of doing business in space is multiplied by the risks of the space environment. This is no trivial footnote in a business plan. While the indirect costs of operating in a primitive and hostile setting are expected to decline over time as settlement commerce grows and matures, there are real risks to consider beyond the usual classic business-related concerns of product, price, place, promotion, and people.

At the top of the list is death. Actually, that's the whole list as far as I'm concerned. In addition to physical death is the professional death of lost or damaged reputation and the psychological death that comes from failure and dreams unfulfilled. All three types of death are highly probable in space. I could add a long list of the different methods death can come in space, but I see no reason to ruin your day. You're welcome.

References

Monopolistic Competition. (2019). *Investopedia*. Retrieved from https://www.investopedia.com/terms/m/monopolisticmarket.asp

12

Visions of Space

Because there are often conflicting views about the importance of space settlement for the future of Earth, and because these views tend to be strongly advocated by one group or another, it is easy to think of these differences as an "either/or" argument. This would be a mistake. The variance of thinking about the purpose and the use of space all share the same passion for the premise: Humankind should go, must go, to space (a) to extend the human footprint in the solar system, (b) to seek new resources to enhance Earth's limited capability to provide for a rapidly growing population, (c) to establish a military presence in support of political and economic objectives and to provide security for partners and stakeholders, (d) to redefine our collective destiny as a space-centric species, and (e) to alleviate global stresses and pressures (environmental, population, industrial, economic, political, etc.) on our planet in order to allow it to heal and return to its natural equilibrium. The arguments about differences lie in the methodology.

There's no Space like Home

Humans have been thinking about our relation to space as far back as prehistory. In its simplest form, Space is the "other" place. An unknown alternate realm where stars and gods live, where myths and origin stories are born. A place of unknowable nature, of incomprehensible reality, of indecipherable meaning. A place not meant for people. This foreboding view has of course changed in the last century. We now see space as part of our natural world, the larger context of our own existence, of who and what we are. And we want to know more about it. More than that, we want to tame it. One of the lessons we learned about ourselves as our species spread across the globe is that we crave control; control of nature, control of other people, and control of ourselves and of our destiny and purpose. This sense of control, of taming the wild frontier of space, is taken for granted in the different visions of space today. There is no vision of space that fully rejects control and adopts a Laissez-faire approach as the means to the end goal of propagating human settlements throughout space. The collective idea is that we are going to space to *DO* something with purpose, not just to *BE* there.

There are arguments that caution future space pioneers to not disrupt the environment on the Moon, on Mars, Europa, etc. These well-intentioned hopes ignore the simple truth that to go to space is to make new frontiers capable of supporting human settlements and commercial ventures. This process will be disruptive and transformative by nature. Just as the Heisenberg Principle suggests that the mere act of observation will alter the subject being observed, so we can expect our human presence in space to alter the places we visit (Jha 2017). Our being in space will change space.

Another common truth about going to space is the act of leaving Earth. Unlike *going* to someplace, *leaving* a former home is often done because of material or spiritual necessity. There is something about the old place that makes restless pioneers seek something new: the lack of success, unsympathetic people, one too many dreams shattered, no more mountains to climb. American pioneers didn't come to California for the surfing; they came because they wanted a better life than they already had. They were leavers. When people begin to emigrate to space, to abandon their life on Earth, what will be the reasons? Famine, poverty, pandemic illness, political persecution? From where we sit in the first quarter of the twenty-first century, the idea of going to space is an adventure, a mythic fantasy of bold exploration. But when reports come back to Earth telling of the hardships suffered on Martian settlements or on lonely outposts in deep space, what will motivate the next wave of space pioneers to be leavers?

Just as there will be leavers, there will be goers. These space wayfarers will be motivated by different desires, a hope of being part of a new chapter in the human story. Some will eagerly go to space out of a sense of humankind's destiny. Others will see the vast space frontier as open range land, unlimited opportunities for wealth and for building a branded reputation in the space creation story. Whether goer or leaver, some emigrants will be ready for the ordeal and others will fail their personal tests outright either by chance or by ill preparation. Both will find common cause with the others who have come to space, for whatever motivation, to take a chance, persist, and make space habitable for greater humanity.

I suspect the reason there are so many visions about the future of space is because everyone has their own idea of the future. Some are optimistic, others not so much. Some are motivated by the science fiction adventures they read, others are fascinated by the science and engineering they've studied. Some want to escape a planet they perceive to be on a death spiral, others see going to the solar system as a logical next step in humanity's manifest destiny. No matter how a person describes their personal vision of space it boils down to an overwhelming agreement that people *will* populate space, that the planets, moons, asteroids, and even the voids between them will become humanity's future homeland. If the subject of space was brought up a half century ago settlements in space would have been only one option on a list that included scientific exploration, military use, and the general social good that comes of exploring our cosmic neighborhood. But now the groupthink *believes* in human settlement of space.

There are a few variants to this collective view, but there is no complete consensus about the timing, the motivation to leave Earth, or the reason to go to another world. In truth, many of the people I've talked with describe their vision of space in the most general terms, often citing broad values like destiny, and economic opportunity. But all admit they don't know what they don't know; that the whole topic of space settlement is fraught with uncertainty and risk. Sometimes the space frontier is described in terms of the opening of the American western frontier. There are parallels, the pioneers slogging through the western wilderness from St. Louis to California or to Oregon risked their lives to claim a new life. But there are also basic differences such as an abundance of food and potable water out on the frontier compared to space which, for now, is strictly a BYO affair. Still, the fundamental image of a pioneer family leaving a bad situation with hopes of creating a new and prosperous life transfers neatly to an ideal of space pilgrims striking out with similar hopes and creating a better future for themselves.

No matter the vision, whether old pioneers who were inspired by "dime novels" glorifying the abundance of the western frontier or modern space settlers encouraged by science fiction thrillers, there are consequences of taking life-altering risks to start over in a new frontier. The promise of independence, the hope of economic gain, the desire for religious freedom, or the wish to escape the crushing formalism and behavioral rules of the modern civilized world in exchange for a life less constrained. All these motives, real or merely imagined, will help cobble a more complete vision of a new life in space.

Here, in abbreviated summary and no specific order, are a few of the dominant visions of space. All the visions described below are well-intentioned—they are born of sincerity and an attempt to cast space in a constructive and positive frame. Some are feasible. Others lean more to the fanciful. But all represent the wide scope of aspirations about future space settlement.

Saving Planet Earth by Making it a Fallow Field

This vision says that the Earth can be saved from itself; the planet's illnesses can be cured. The idea is for people to remove themselves from our planet to let it lie fallow, regenerate, and heal. All that's needed, advocates of this vision maintain, is a respite from human influence for a generation or two so that the destructive influence of human civilization (e.g., overfishing the oceans, over production of diminishing crop lands, covering the planet in concrete, industrial pollution, etc.) can be reduced in time. Unknown is how long it will take for the planet's health to get back on track. The process of reversing the negative impact started by the industrial revolution may take a lot longer than anyone can imagine because of all the plastic in the ocean, mass extinctions of animal species vital to the world's ecology, and other consequences of industrialization.

A note of optimism for this idea: when the planet essentially shut down due to the 2020 covid-19 pandemic, it didn't take long for smog and other pollution to clear from the skies of Beijing and Los Angeles. The pandemic hinted at a proof of the concept that the Earth can heal itself. Can our planet be healed without evicting a significant portion of human population to space? But, wait! What if the world came together in a near-unanimous accord and instituted global reductions of pollution, greater control of carbon emissions, etc.…? Oh, wait, that's already been suggested without much success.

"Plan-B" Strategy: Abandoning Planet Earth

Plan B is like the Fallow Field vision except that the mass exodus of people to space is a one-way trip: people never return. This vision frames space as a pandemic insurance policy where space settlements are the lifeboats in which humankind can flee to escape the death spiral decline of the planet, in case of widespread disease, catastrophic climate change, famine, rampant wildfires, vanishing arable croplands, nuclear holocaust, or any other life-extinguishing disaster that could wipe out humanity. In this vision, space is where you go when there's no other way to fix problems on Earth. Space is where to go in order to save the human species and restart human civilization all over again. Elon Musk and others advocate for this solution.

Space Settlement Is Humanity's Manifest Destiny

This vision is straightforward. Simply put, humankind is going to space because that's what our species is meant to do. Along with this line of thinking is the conceited notion that we will soon be masters of the solar system just as we have become masters of our original home planet (there is a bit of circular logic at play given we supposedly will have no rivals in this competition). A justification for man's manifest destiny in space comes from the adventurous side of our nature that says we are curious beings who thrive on discovering new frontiers and will forever set out to discover what's just over the horizon. When Sir Edmond Hillary was asked why he wanted to be the first to climb Mt. Everest he is reported to have said, "because it's there." This is much the same thing.

This romantic notion about pursuing our ancient hunter-gatherer nature seems to have a degree of appeal. But if the urge to trek across the empty tundra in search of "what's just over the horizon" was all-consuming then why did our species give up all that trekking around the wilderness to grow crops, build cities, and trade goods?

Military/Political Strategic High Ground

As of this writing there are dozens of nations and corporations seriously looking to establish permanent colonies, outposts, military bases, or settlements in space by mid-twenty-first century or sooner. The energy and momentum of this goal has substantial financial backing and political will. The establishment of new colonial expansionism in space will leverage the expected vast raw resources in space to generate profit and power for those who elect to exploit space's riches.

Along with the obvious motivation of greed ("Greed is good," preached deal-maker Gordon Gecko in the 1987 film, *Wall Street*) is the potential greenfield advantage of expanding political power in space. Dominant nations on Earth have the advantage of funding joint military and commercial outposts that will extend their political influence in space as well as establishing future settlement communities. Colonial expansion by corporate-government partnerships could carve out lines of power and influence in space that will likely have a lasting effect for generations.

Theological/Spiritual Rationalization: Being One with the Force

Some consider space as an evangelical opportunity. Others see the prospect of space settlement as a utopian chance to create a cloistered community for fellow members of their religious group. As discussed earlier, the American frontier presents a rich history of religious sects and cults seeking isolation from the "corrupt" civilized world. The social revolution of the 1960s saw the rebirth of this movement in the form of ashrams and communes where people seeking an alternative spiritual life could go and explore their purpose in life. For the spiritually opportunistic entrepreneur, space may provide seekers with a chance to find their personal purpose or have other "cosmic" spiritual encounters in the form of specialized space tourism experiences. Meditating on the Moon may be the new revelatory Esalen Institute weekend experience.

Social and Political Experimentation and Social Diversity

Recalling a primary theme of Frederick Jackson Turner's *Frontier Thesis*, the closing of the American frontier was, for him, a signal that there were no longer any safe havens for social outliers. Instead of escaping to the frontier wilderness to avoid contact with the civilized world, those who self-selected to opt out of civilized society were now forced to integrate and participate in some way with the rule-heavy bureaucratic formalization of society. We used to call these people names like hermits or mountain men. But now, the opening of new space frontiers will unencumber those individuals who wish to live apart from what they consider to be an oppressive and restrictive social structure. Not because they have a religious or spiritual agenda but because they feel they just don't want to be part of the structured (i.e., civilized) world.

Economic Pragmatism and Expansion

This view of space is one of commercialization. It positions itself as the backbone of the future of space; the future fuel that will drive the industrialization of the solar system. Corporate planners, entrepreneurs, investors, and national interests see the abundant resources and raw materials in space as a ready-made platform for tremendous economic expansion. As discussed earlier, whether Earth becomes the primary customer and trading partner of industrial space, or whether space business primarily serves space-based customers, there will be dramatic economic growth and development in space. Along the way, space settlements will be established as villages and towns on the frontier; hosts to business ventures big and small. As these commerce centers gain momentum, they will likely attract more settlers and more business ventures by the consequence of their financial dealings. In due time, maybe a century of two, the space economy will take on a life of its own and likely eclipse the global economy of Earth.

Mission Statement Matchup

Thanks to expensive management consultants, every company has a well-scrubbed vision statement, mission statement, values statement, and branding statement. (These statements are typically posted in the employee break room along with the map showing the quickest way to flee the building in case of fire.) In most cases, one company's set of statements is much like another's, homogenized and tepid. This is partly due to institutional laziness. This may be on purpose, because well-established company political culture dictates that mission/vision/values be closely aligned with their customers and competitors. It's better to be part of the herd than an outlier.

In an emerging business sector, like the new space sector, most companies are still outwardly wedded to the founder's sense of idealism and purpose. Founders of breakout start-up firms often parlay their image as an iconoclastic rebel as an attribute of the brand's innovative image: a fearless intrepid individual unafraid to stray from the herd and boldly venture into uncharted commercial territory. This image, along with a wardrobe consisting of jeans and a black T-shirt, sends the message to potential investors that s/he is no ordinary leader. Thus, the company is no ordinary company as is further evidenced by the firm's mission statement. As an example, here are the mission statements from three prominent space firms: Blue Origin, SpaceX, and Virgin Galactic. These statements are taken directly from the company website.

- **Blue Origin.** "Blue's vision is a future where millions of people are living and working in space. In order to preserve Earth, our home, for our grandchildren's grandchildren, we must go to space to tap its unlimited resources and energy. If we can lower the cost of access to space with reusable launch vehicles, we can all enable this dynamic future for humanity."

- **SpaceX**. "SpaceX designs, manufactures and launches advanced rockets and spacecraft. The company was founded in 2002 to revolutionize space technology, with the ultimate goal of enabling people to live on other planets."
- **Virgin Galactic**. "Our mission, to be the Spaceline for Earth, means we focus on using space for good while delivering an unparalleled customer experience."

There are some interesting discriminators at work in these statements. Virgin Galactic is the only one of the three firms that claims to be very much about creating a consumer brand that provides a top "customer experience." This is a laudatory corporate objective, something that can be measured and improved upon over time. But it falls short of specifics about what a "Spaceline" is (although it was recently revealed that Virgin plans to apply the technology developed for suborbital space tourism flights to also provide point-to-point transportation around the world).

SpaceX is clear that it sees itself as a producer of space transportation hardware. Nowhere is there mention about enhancing the customer experience or the long-term impact of space transportation on the greater global community. With this mission statement they have positioned themselves as a utility provider to serve other organizations—both public and private—as a low-cost/high-quality enabler of space transportation technology. Unlike Elon Musk's other ventures like PayPal and Tesla which directly face the consumer, SpaceX employs a B2B business model.

Blue Origin has the courage to move beyond the transactional branding efforts of SpaceX and Virgin Galactic with transformational references to altruistic values like preserving Earth for "our grandchildren's grandchildren." This hopeful tone of long-term commitment claims a purpose more about community service than technical achievement or commercial growth. Blue doesn't ignore the business side of the ledger and emphasizes the goal of "lower cost access to space" as a mechanism that will "enable the dynamic future for humanity."

When viewed through a simpler, distant lens, these three firms are merely space companies. Dive a little deeper and it becomes apparent that organizations in the new space sector are not monolithic; that they have different visions of their purpose and how they measure success, that the space sector is already developing complexity and differentiation.

Plan Blue and Plan Dragon

There are two dominant narrative propositions that make a case why humans should settle space. Let's call them Plan Blue and Plan Dragon for purposes of our discussion. These visions or philosophies, at their core, are what a skilled marketing professional would call value propositions; the selling points a peddler makes to convince a buyer why the product is (a) intrinsically important (a car will provide you with convenient transportation), (b) better than the competition (car X has better technology, gas mileage, safety, sex appeal, etc., than car Y), and (c) a perfect match for your lifestyle and personal interests (car X is the right car for your growing family for outings on the weekends and for you to make an impression when you entertain your boss after work). There are other cases for settling space, but these two perspectives tend to take up most of the oxygen in the current discussion.

Both Plan Blue and Plan Dragon have their supporters and their detractors. Both plans decry the faulty logic and unethical appeal to emotions employed by their competitor. Both plans are based upon core truths about the impact of humankind's abuse of our planet's resources and the need to alter the destructive path we are on for the sake of "our grandchildren's grandchildren." And both plans wind up at the same place: a civilized solar system taking full advantage of endless mineral riches for the betterment of human civilization.

There are differences, of course. For example, Plan Blue prefers outposts constructed in space to take full control of the full bio-environment. Plan Dragon thinks settlements should be located on a planet surface to replicate our "planet chauvinist" notions about proper human habitats. But beyond these simple orthodoxies both plans see opportunities for the emergence of new social structures, for advanced economic development, and for generations of humans with dwindling allegiances to their historic origin: Earth. These special visions are worth discussing.

Plan BLUE: Jeff Bezos' Vision

During an interview about his objective for developing space capabilities in early 2019, Jeff Bezos spoke candidly about his long-term vision and his continued motivation for investing at least one billion dollars a year in his Blue Origin enterprise. "I want to take the assets that I have from Amazon and translate that into the heavy-lifting infrastructure that will [help] the next generation to have dynamic entrepreneurialism in space—kind of build that transportation network. That's what's going on, that's what Blue Origin's mission is," said Bezos (Mosher 2019).

To get the process started Bezos believes that developing infrastructure and transportation capabilities are key components. He cited the lessons learned from launching his company, Amazon. "I started Amazon in my garage

24 years ago—drove packages to the post office myself. Today we have 600,000-plus people, millions and millions of customers, a very large company. How did that happen in such a short period of time? It happened because we didn't have to do any of the heavy lifting. All of the heavy-lifting infrastructure was already in place for it."

Bezos sees his mission as an optimistic way to preserve and heal the world, not as a pessimistic escape plan of building lifeboats to abandon a dying planet.

"We want to go to space to protect this planet. That's why the company's named Blue Origin—it's the blue planet that's where we're from. But we also don't want to face a civilization of stasis, and that is the real issue if we just stay on this planet—that's the long-term issue." His plan is to settle space to expand human habitation in the solar system not as an alternative but as an extension of our human migration beyond our origin planet. In so doing, Jeff sees the industrialization of space as a way to help heal our mis-managed ecosystems, a way to let the Earth heal by transferring destructive industrial practices off planet. Part of his plan is to utilize space for heavy industry, energy generation, and other commercial activities that impact Earth's environment. He anticipates that this would help to alleviate human suffering caused by scarcity of vital goods, food shortages, and rationing caused by population growth.

Another goal of settlement is to create new opportunities for entrepreneurs in space. "Once you get into space, you can really unleash a lot of creativity." In this way space settlement and economic development go together. Some of his optimism about the future of human space settlement comes from his belief that, "The solar system can support a trillion humans, and then we'd have 1000 Mozarts, and 1000 Einsteins. Think how incredible and dynamic that civilization will be." This potentially fits with Bezos' Amazon business model where independent entrepreneurs are hosted on his site and profits shared. Although unsaid, it is possible that Jeff will create a commercial infrastructure that will encourage small independent commercial ventures to piggyback on his more robust organization.

Planet Chauvinists

As to the logistics of his future vision, Jeff acknowledges his strong influence by Gerard K. O'Neill who proposed mixed-use habitats in his book, *The High Frontier* (1976). The habitats would be constructed in space where settlers would live and work in giant wheel-shaped structures that would induce artificial gravity by their constant rotation. Bezos explained that there is no need to be "planet chauvinists" and just build settlements on the surfaces of worlds in the solar system. Instead, he thinks we can build our own habitats wherever it is best for us to live, work, or explore our solar system.

An example of a vital infrastructure component that would support Plan Blue is energy generation. Dr. Greg Autry, the director of the Southern California Commercial Spaceflight Initiative at the University of Southern California (USC), and vice president at the National Space Society, has researched the commercial uses of space. He observed that, "Sustainable energy advocates in the U.S. military and the Chinese government are actively pursuing space-based solar power, but just making solar cells damages the environment due to the caustic chemicals employed." Solar power technology can be manufactured in space from resources found in space. "[Space-based manufacturing] will reduce the damage caused by manufacturing and mining on Earth" (Autry 2019). Shifting toxic and hazardous industrial activities off-planet is a key element of Plan Blue.

Plan Dragon: Elon Musk's Vision

Unlike the more altruistic vision of Jeff Bezos based on a belief that settling space is a way to save our planet by moving destructive industrial activities to space among other tactics, advocates of Plan Dragon, like Elon Musk and others, believe that space settlement is how humanity will escape the harmful trends already in motion on our planet by building a new self-sustaining society on Mars so we can leave our old planet behind.

Plan Dragon advocates take the "glass-half-empty" view of Earth's future. They believe that things aren't going well for our planet, either because of human mismanagement or because of the natural order of Earth's continual evolution, and that space settlement is a solution ready-made to ensure the human species doesn't end up extinct like the dinosaurs or the 8-track tape deck. Whether due to cataclysmic disasters like giant asteroids and massive volcanic eruptions, billionaire entrepreneur Elon Musk is one who is very worried (and very vocal) about the next inevitable apocalypse. For Musk and other Plan Dragon believers it's not a matter of if the Earth becomes uninhabitable but a matter of when we should abandon ship. Supporters of Plan Dragon are so concerned about the imminent destruction of our planet that they believe we need to get off Earth and become a multi-planet species as quickly as possible. Musk's plan is to relocate one million people to Mars by 2050 using his Starship rockets (Mosher 2015).

Their reasoning is simple. If the chances of Earth's apocalyptic demise are high, then it makes sense to prepare for mass extinctions caused by the next giant asteroid, famine, social unrest, or pandemic by evacuating to settlements in space. One flaw in this thinking is if a cataclysmic event takes place, like a nearby star exploding, then there will be no safe place to hide; we will all likely get vaporized. Another flaw is that the killer asteroid, pandemic, or other fatal event may just as likely happen to Mars as it may happen to Earth. This could be mitigated by sending many millions of people to many thousands of new outposts in the solar system. This would vastly increase the chances of our species' survival. And it would also require the purchase of lots of SpaceX rockets. So it goes.

Musk says we can't afford to wait around and find out. He and other Plan Dragon thinkers have a high sense of urgency about planning for the worst, like modern-day "preppers" who stock up on dried beans and Spam just in case we're invaded by Canada, the Plan Dragon folks say that the sooner we establish colonies and settlements on Mars and other rocks in our solar system, the better chance we have to stave off human extinction. The Plan Dragon plan is to relocate one million people to Mars by mid-century in hopes that their industry and genetic diversity would help to perpetuate humanity. Others are less optimistic and estimate making Mars fully habitable will take many generations (Robitzski 2019).

Plan Z: Permanent Staycation on Earth?

Whether optimistic or pessimistic about the future of human life on Earth, we are all going to be stuck here for a long time. Establishing a colony on Mars doesn't mean the problems of space radiation, the impact of diminished gravity on the human body, or providing basic nutrition and other necessities of life, will be solved all at once. For a significant period—some say generations—space outposts and settlements may remain dependent on Earth for survival. Relying on Earth means the hope that space settlement will lead to space autonomy is incorrect. Dependence on Earth will tether early attempts at settlement autonomy to Earth for a long time.

Although the reasons why we should go to space are often profoundly different, they have many common threads that stitch them together. Advocates for space settlement dream of a future in space with optimism and hope. Even pessimists who believe Earth is on a collision course with catastrophes like overpopulation, famine, pandemics, environmental collapse, and social breakdown see the prospect of settling space optimistically as a second chance to preserve humanity. Or maybe the real vision of space is "just the stuff that dreams are made of" (Dashiell Hammett, *The Maltese Falcon,* 1941).

References

Autry, G. (2019, July 20). Space research can save the planet—again. Foreign Policy. Retrieved from https://foreignpolicy.com/2019/07/20/space-research-can-save-the-planet-again-climate-change-environment/

Jha, A. (2017, February 22). What is Heisenberg's uncertainty principle? *The Guardian*. Retrieved from https://www.theguardian.com/science/2013/nov/10/what-is-heisenbergs-uncertainty-principle

Mosher, D. (2015, October 10). *Elon Musk: We need to leave Earth as soon as possible*. Business Insider. Retrieved from https://www.businessinsider.com/elon-musk-mars-colonies-human-survival-2015-10

Mosher, D. (2019, February 23). *Jeff Bezos just gave a private talk in New York. From utopian space colonies to dissing Elon Musk's Martian dream, here are the most notable things he said*. Business Insider. Retrieved from https://www.businessinsider.com/jeff-bezos-blue-origin-wings-club-presentation-transcript-2019-2#reusable-launch-systems-arent-good-enough-alone-bezos-said-they-have-to-be-easy-to-reuse-or-costs-get-too-high-negating-the-point-of-their-existence-2

O'Neill, G. K. (1976). *The high frontier: Human colonies in space*. Mojave: Space Studies Institute Press.

Robitzski, D. (2019, July 30). Reality check: It would take thousands of years to colonize Mars. *Futurism*. Retrieved from https://futurism.com/the-byte/thousands-years-colonize-mars

13

The Space Economy Is Already Here

If you're waiting for the start of the new space economy, then this is your wake-up call. Commercial space ventures have been open for business for a long time. Space enterprises are engaged in a broad scope of profitable ventures with new start-ups coming online all the time. Some ventures provide technical equipment, others focus on delivering services like data analytics or consumer communications. Not only has the new space economy already taken root in the industrialized world, its influence is also spreading across national boundaries into emerging markets and redefining itself across the fuzzy demarcations of traditional business sectors. It has slowly spread like a stealthy space invasion, hiding in plain sight, revealing itself most prominently when events like Mars missions and scientific probes grab the spotlight of public attention.

The cosmos economy is a lot more than building cities in space and industrializing the solar system. It is also economic development on Earth to support a tsunami of new space ventures. It is global economic development opportunities, it is the emergence of a brand-new business sector, it is technical creativity and innovative technologies originally developed for space and then implemented here on Earth. It is greater opportunities for employment and for building lifelong careers in new professional fields. It is bold leadership and strategic visions of how the new space economy will transform the business on Earth in the coming centuries (Accion Systems 2020). The space economy isn't something that will take place far, far away "out there," because the space economy is already firmly grounded in the technical and financial capabilities of today's visionary entrepreneurs.

The space economy wasn't always called the space economy, it was a subset of the aerospace and defense industrial complex. In the beginning, it was a collection of programs, projects, and missions run by major prime contractors, affiliated companies, subcontractors, and the supply chains that fed the production of space hardware like rockets, satellites, and other mission-related technologies that employed the engineers, technicians, and administrators who worked in the new space sector. Current commercial space activities like communications satellites, data collection, data analytics, space tourism, and others are the result of decades of development.

Space Is a Private Affair

The big difference about today's space economy is that private space companies, companies whose mission and purpose is exclusively space, are taking the lead and setting the priorities based upon their view of the market. New private space companies are springing up all around the globe as are national space agencies. Gone are the days of a space race dominated by just two superpowers vying for national prestige; now space is accepted as a new opportunity for economic development, a chance to participate in the early stages of a new economic force. The intersection of public space programs and private space companies has shifted control of space from government-backed space agencies like NASA to private providers like Blue Origin and SpaceX seeking to set their own agenda and who wish to establish their own priorities about space sector economic development. Instead of putting NASA projects at the top of the priority list, private companies have taken the upper hand and established where space activities are going based upon their own business strategies and profitability targets. They have also set a more rapid pace of implementation that has proven cheaper and faster to market than the old bureaucratic process typified by NASA. Professor Matthew Weinzierl of Harvard Business School observed, "[Private space companies] have their own reasons to go to space, and then NASA is trying to leverage what they can do, what the private companies can do, to better achieve its own goals. It's really about not so much a shift in the scale of commercial activities in space but a shift in the nature of the relationship between the public and private sector" (2018).

There is a convergence of thinking about the role of private space companies. In another paper Dr. Weinzierl observed, "The shift from public to private priorities in space is especially significant because a widely shared goal among commercial space's leaders is the achievement of a large-scale, mainly self-sufficient, developed space economy" (2016). The lure of the new space sector has attracted a range of comments about its full potential. Professor Weinzierl concluded that, "... Neil deGrasse Tyson and Peter Diamandis have been credited for stating that Earth's first trillionaire will be an asteroid miner." (This could be a tricky calculation. The oversupply of a precious raw material [e.g., gold] could instantly turn that valuable scarce material into a relatively worthless commodity if there is a limitless supply. The high cost of mining the material could instantly erase the benefits of mining the once-precious ore.)

Looking back on the developments of the last half century Daniel S. Goldin, NASA administrator from 1992 to 2001 said, "The commercial space industry, built on significant private investment, is providing thought leadership … to help define new directions for America's space program. The question for us earthlings is: Do we have the vision, technology and tenacity to build commercial operations to use these natural resources in our neighborhood, or are we living in a field of dreams?" (2019).

The impact of new $415B space economy (Bryce 2020) is already being felt in a broad range of industries not associated with space. Today commercial space is weather forecasting, GPS navigation, new materials for manufacturing, new production processes, advanced global communications, healthcare and pharmaceutical innovation, and even precision crop monitoring for farms. Technical and economic benefits flowing to Earth have grown over the last several decades and will continue to grow as more private companies and nations recognize the full economic potential of space. The space economy isn't something brand new, it's been here for decades, steadily developing new technologies and creating new markets. Hiding in plain sight embedded with the rest of the global industrial economy.

A major benefit of the space economy is the new opportunities for employment. This new economic sector is a greenfield for women and minorities for decades to come. The space economy offers wonderful opportunities for new start-ups, entrepreneurial innovations, and, most importantly, increasing job growth for women and minorities who wish to launch or transition their career to a global growing industry. Space companies large and small employ not just engineers, they also need frontline workers, managers, and a full range of well-paid positions.

The Current State of Commercial Space

Commercial space ventures have spawned venture capital (VC) firms that fund new space start-up ventures, Space Angles and SpaceFund are two examples. Both are growing and tracking significant portfolios of new space enterprises. Their business model is to service other commercial space firms. Here is how Meagan Crawford, Managing Partner of SpaceFund, describes her long-term strategic view as being an "empresario" of space start-ups.

> We're looking at the entire in-space supply chain and how it's going to develop in the next 10 years (the life of the fund) and are working to put all the pieces together. We're very much trying to ensure that our portfolio of companies is complementary, and whenever possible are working together to help each other. While we haven't yet been directly involved in starting new companies to fill holes in the supply chain, it's definitely something we've talked about. (Email exchange, June 19, 2020)

The goal of the venture capital (VC) firm, a business that recruits investors to fund start-up firms in their portfolio, is to show profitability in the first 5–7 years that will produce a profitable exit. An exit can be either the acquisition of the start-up by another firm or an initial public offering (IPO) on a public stock exchange like Nasdaq or the Dow Jones. Funders know that the chance of this happening for even a mainstream sector like technology or health care is slim and for a brand-new investment category like space the chances of success in such a short time are even less. But space is different because the potential rewards could well be worth the risks of investing.

Note, also, that most entrepreneurial firms seeking funding have a business plan that aligns to the current commercial space profile, that is, launch vehicles, sensors and data collection in space, data analytics, booking and managing payloads bound for the ISS, and other businesses that cater to low Earth orbit (LEO) space. Few of today's VC undertakings take the longer view aimed at serving a mid-century space economy built on active Lunar and Martian outposts with all the ancillary materials and supporting services needed to establish and maintain these permanent commercial operations. Yet, investment momentum in the commercial space sector is picking up and the highly anticipated date of 2045 when most predict a critical inflection point on the space industry growth curve is just around the corner.

Because all today's customers for commercial space are still down on Earth the types of services (and space commerce today is mostly services) are tailored to meet the demand of Earth-based businesses: Data collection from space that uses sophisticated sensors to capture a daily detailed photographic survey of the planet, data analytics which spin-dries the raw data collected by satellites into usable information for a growing variety of industries, the global positioning system (GPS) which aids navigation activities for industries like shipping and farming (crop maintenance), communications and entertainment providers like Direct TV, and space tourism which has seen Virgin Galactic listed on the New York Stock Exchange to great success even before a single dollar of revenue was realized.

Some examples of commercial space ventures include SpaceBridge Logistics which acts as an intermediary broker of excess rocket capacity by helping satellite customers book the best rocket to fit their payload specifications. The company also provides satellite customers with the capability of a web interface to "buy a rocket" tailored to their specific launch preferences. NanoRacks provides similar rocket booking services and uniquely provides companies that have experiments in space on the International Space Station (ISS) regular monitoring of the customer's research progress. Made in Space operated the First 3D printer in space and uses their additive manufacturing system to build products that are best produced in micro-gravity like ultra-high-quality ZBLAN optical fiber that has superior data transmission due to near-flawless characteristics because it is manufactured in zero gravity.

The marque commercial space activities are rockets (SpaceX, United Launch Alliance, and others) along with launch and mission support services, and satellite production. SpaceX has decreased costs by developing the ability to reclaim and reuse the first stage of their rockets. This cost reduction has dramatically dropped the cost per kilogram of payloads which has made them highly competitive when bidding for launch contracts. Northrup Grumman (NGC) has entered the commercial space arena with a solution to prolong the life of an expensive satellite that has exhausted its fuel and the ability to maneuver in orbit. The NGC remedy attaches a mate to a satellite with a new fuel supply, thus giving a second life to a fully depreciated technology investment. Bigelow Aerospace Corporation has developed inflatable habitats designed to function in space and on a planet surface. The Bigelow Expandable Activity Module (BEAM) habitats are designed to withstand micro-meteorite strikes and mitigate deadly space radiation. They have been attached to the Tranquility module of the International Space Station (ISS) where they were successfully tested. These and other commercial ventures into near space are not only generating revenue but are legitimizing the industrial space sector. As more and more firms look to space for new market opportunities, the success of these early market entrants will entice others to follow.

New Space Investors

Initially private space was the domain of billionaires with very deep pockets who were willing to invest great amounts of capital from their private accounts.

Bezos, Branson, Bigelow, and Musk come to mind. Reporter Christian Davenport explored the role of billionaire space investors in a *Washington Post* article (2018) and found that most of the high rollers were more focused on space than profit. Now, with their earlier success demonstrating the viability of the space market, other private investors are backing new space ventures with hopes that they will have an early edge in the space sector.

The following three space-oriented VC firms are included to show some of the ventures they've funded in order to paint a general picture of commercial space today. This snapshot is not a comprehensive list of space start-ups. Given the fluid nature of the space industry some of the start-up companies listed here may no longer be operating as viable organizations or may no longer have a relationship with the VC indicated below. Some companies are funded by multiple investment firms. The brief company descriptions are from their websites.

- **Space Angels Network.** https://www.spaceangels.com/

    - **Analytical Space.** https://www.analyticalspace.com/ (Analytical Space 2020) "Analytical Space, Inc. operates a satellite relay network that stores and provides access to data for satellite operators. It offers backward compatible, throughput, and latency services. The company was founded in 2016 and is based in Cambridge, Massachusetts."
    - **Astrobotic.** https://www.astrobotic.com/ (Astrobotic 2020) "Astrobotic Technology, Inc. is a space robotics company that seeks to make space accessible to the world. The company's lunar lander, Peregrine, delivers payloads to the Moon for companies, governments, universities, non-profits, and individuals."
    - **Kepler Comm.** https://www.keplercommunications.com/ (Kepler 2020) "Kepler's founding purpose is to build infrastructure for space-based connectivity. Our infrastructure will connect launch vehicles, space stations, habitats, and other satellites. We want to bring the Internet to space!"
    - **NanoRacks.** https://nanoracks.com/ (Nanoracks 2020) "Nanoracks is both the largest commercial user and private investor on the International Space Station (ISS). With customers from 30+ nations around the world, Nanoracks offers launch services on the International Space Station, suborbital space, polar orbit, and provides everything necessary to place payloads into space."
    - **Planetary Resources.** https://www.planetaryresources.com/ "Planetary Resources was acquired by ConsenSys in October 2018. Planetary Resources focused on space mining: harnessing the natural resources of

celestial bodies, from spacecraft design to tools for technical and economic analysis of resources throughout our solar system."
  - **World View**. https://worldview.space/ "World View [uses] un-crewed [inflatable] Stratollite flight vehicles to collect data at lower altitudes than satellites in the stratosphere, navigates using winds at different altitudes, and offers persistent observation of areas of interest for days, weeks, and months at a time."

- **Founders Fund**. https://foundersfund.com/ (Founders Fund 2020)

  - **Accion Systems**. https://www.accion-systems.com/ "Developed in-space propulsion technologies that optimize scalability, performance and efficiency. Accion's flagship product, Tiled Ionic Liquid Electrospray *(TILE)*, uses proprietary electrospray thrusters, bringing electric propulsion to satellites of all sizes."
  - **Moon Express**. http://moonexpress.com/ (Moon Express 2020) "Our MX family of flexible, scalable robotic explorers are capable of reaching the Moon and other solar system destinations from Earth orbit. The MX spacecraft architecture supports multiple applications, including delivery of scientific and commercial payloads to the Moon at low cost using a rideshare model, or charter science expeditions to distant worlds."
  - **Planet Labs**. https://www.planet.com/ Monitors areas of interest, validates information on the ground, and determines trends relevant to the client's industry. Planet employs a fleet of proprietary satellites to collect data and provide geospatial analytic insights at the speed of change that enable organizations with the data necessary to make informed, timely decisions.
  - **SpaceX**. https://www.spacex.com/ "SpaceX's family of launch vehicles and spacecraft were designed from the beginning to take humans to Earth orbit, the Moon, Mars and beyond."

- **SpaceFund**. https://spacefund.com/

  - **Made in Space**. https://madeinspace.us/ (Made In Space 2017) "First 3D printer in space, additive manufacturing system now used under the ocean and in Antarctica. Manufactures ultra-high-quality fiber optics in space." [Note: Made in Space was acquired by Redwire in June, 2020.]
  - **Axiom Space**. https://www.axiomspace.com/ (Axiom Space 2020) "Commercial space station designed to evolve from International Space Station both in technology and customer base/income generation. First

systems scheduled to fly with first paid commercial 'spaceflight participants' on the ISS in early as 2020s."
- **Orbit Fab**. https://www.orbitfab.space/ (Orbit Fab 2020) "Satellite re-fueling, propellant storage and transfer technologies. Flew on International Space Station, demonstrated transfers. Several patented technologies including docking clamp and fill/drain port for satellite refueling."

Exit to Profitability

As an indication of the robust nature of space investment, Voyager Space Holdings (2020), led by CEO and Chairman Dylan Taylor, provides proven space ventures with an exit opportunity; the next step for entrepreneurs wishing to convert their shares to cash. Voyager's business model is to assemble a balanced portfolio of complementary space companies with (a) proven success based upon ability to execute and to serve customers in their market space, (b) seasoned management team with balanced functional skills who work with a sense of urgency to achieve their core vision, and (c) financial acumen. Taylor outlines his picture of the future of the space economy as a sector with increasing unmet demand and with more and more global cross-border companies collaborating to achieve common interests. He decries the efforts of, "too many companies attempting to enter the space sector based more on hype than solid business practices," yet he predicts there will be increased IPO activity in response to unmet retail investor demand for equities of space-related firms. Finally, he lauds the proposed Artemis Accords designed to establish a common space policy platform for nations and private companies, as an opportunity to establish common values and technical standardization.

The increased interest in the financial returns of the future space industry continues to attract investment funds to space start-ups and venture capital firms. Bryce Space and Technology, a consulting firm that tracks and reports on the commercial space sector, reported that 2019 saw investments of more than $2.5 billion for start-up space ventures, the third year in a row showing noteworthy commitments of private investment.

Artemis is the next giant leap to the moon. Named for the twin sister of Apollo in Greek mythology, the Artemis program goal is to establish a permanent human presence on the Moon unlike the Apollo program of the last century that only planned to briefly visit and then a quick return. An orbiting Lunar station, called Gateway, will orbit the Moon and provide logistical support and research for Lunar excursions to the Artemis base. Eventually Gateway will also serve as a staging platform for Mars excursions. NASA Administrator Jim Bridenstine said, "Among other things, [Artemis] will allow us to accelerate our development of the Space Launch System and Orion [the crew exploration vehicle], it will support the development of a human lunar landing system and it will support precursor capabilities on the lunar surface, including increased robotic exploration of the moon's polar region" (2020). Gateway is designed to also assume a key role staging other manned expeditions to Mars, the asteroid belt, and other points of interest in the solar system. The current Gateway configuration will be only 1942 cubic feet of livable space, compared to 13,696 cubic feet on the International Space Station (ISS), and is expected to be occupied by three to four astronauts for just 90 days at a time unlike the ISS tour of duty which can last from 6 months to a year. For some time, there was a policy debate regarding establishing a base on the Moon or on Mars. Implementing the Lunar orbiting Gateway as an intermediary staging platform allows for greater flexibility and will enable expeditions to both celestial destinations.

Artemis and Gateway are much more than scientific programs. They are a firm stepping-stone for establishing the critical technologies necessary to create a commercial space infrastructure. Commerce Secretary Wilbur Ross, interviewed by CNBC reporter Morgan Brennan, sees a clear path from today's Artemis Lunar program to commercial space ventures. He explained the importance of establishing a permanent presence on the Moon that also includes the Gateway space station in Lunar orbit that will act as a staging platform for forays to Mars, the asteroid belt, and other destinations in the solar system. Secretary Ross noted the range of commercial ventures from asteroid (and planet surface) mining, manufacturing, tourism, along with the scope of suppliers that would support this effort. He further predicted the space economy would, "… rapidly achieve a multi-trillion-dollar value within the next few decades."

References

Accion Systems—A New Ion Engine. (2020). *Accion systems—A new ion engine*. Retrieved from https://www.accion-systems.com/
Analytical Space. (2020). *Analytical space*. Retrieved from https://www.analyticalspace.com/
Astrobotic. (2020). Retrieved from www.astrobotic.com. https://www.astrobotic.com/

Axiom Space—The World's First Commercial Space Station. (2020). *Axiomspace*. Retrieved from https://www.axiomspace.com/

Bryce Space and Technology (Carissa Christensen, Janice Starzyk, Simon Potter, & Nick Boensch). (2020). *Start-up space; Update on Investment in Commercial Space Ventures*.

Davenport, C. (2018, March 11). *Why billionaires keep pouring money into the space industry*. Washington Post. Retrieved from https://www.washingtonpost.com/news/the-switch/wp/2018/03/08/why-billionaires-keep-pouring-money-into-the-space-industry/?noredirect=on&utm_term=.d90b7aed93e2

Founders Fund. (2020). *Founders Fund*. Retrieved from https://foundersfund.com/

Goldin, D. S. (2019, July 12). *Preparing humanity for life among the stars*. Aviationweek.com. Retrieved from https://aviationweek.com/defense-space/space/opinion-preparing-humanity-life-among-stars?Issue=AW-05_20190716_AW-05_45&NL=AW-05&cl=article_2&elq2=0fc2f9d13e434c65a844387bdeb091e0&sfvc4enews=42&utm_campaign=20402&utm_medium=email&utm_rid=CPEN1000002537211

Kepler. (2020). Kepler. Retrieved from https://www.keplercommunications.com/

Made In Space. (2017). *Made in space*. Retrieved from https://madeinspace.us/

Moon Express Incl Redefine Possible. (2020). *Moon Express*. Retrieved from http://moonexpress.com/

Nanoracks|Your Portal to Space. (2020). Nanoracks.com. Retrieved from https://nanoracks.com/

Orbit Fab. (2020). *Orbit Fab*. Retrieved from https://www.orbitfab.space/

Voyager Space Holdings. (2020). *Voyager Space Holdings*. Retrieved from https://voyagerspaceholdings.com/

Weinzierl, M. (2016, May 11). *Who owns space? HBS working knowledge*. Retrieved from https://hbswk.hbs.edu/item/who-owns-space-podcast

Weinzierl, M. C. (2018). *Space, the final economic frontier*. Retrieved from www.hbs.edu. https://www.hbs.edu/faculty/Pages/item.aspx?num=54487

14

The New Space Merchants

James Causey (2019), Executive Director of SpaceCom, famously stated that, "Every company is a space company, some just don't know it yet!" (November 20, 2019). This of course implies that every customer of every company is (or soon will be) a de facto space customer. That's all of us! While this is yet to take hold in the general business mindset it is well worth considering. If the tide turns to space-based manufacturing production, mining, agricultural production, pharmaceuticals, space tourism, and a thousand other endeavors, then all of us in the capitalist value chain are now (or soon will be) contributors to and consumers of the new space economy.

The reason commercial interest in space has skyrocketed is because the rules of the game shifted from exclusive government control to open partnerships with private firms seeking a profitable return on their space investments. The old standard operating procedure (SOP) was fully dependent on government funding; going to space was the domain of nations who had the financial resources and infrastructure to develop technologies, launch payloads into orbit or deep space. These space nations (e.g., Russia, the U.S., Japan, China) set the agenda for scientific exploration, military and defense activities, and then took the reputational rewards. Because government policy dictated which programs would receive funding, most companies in the private sector, with a few exceptions, were relegated to supplying components and service instead of creating entrepreneurial commercial ventures. This was not a business-friendly business model.

Central planning, not organic market forces, tends to strangle new ideas and perpetuate the status quo. (The Soviet Union famously kept turning out steam locomotives late into the twentieth century because no one thought to switch the perpetual five-year plan to making diesel train engines.) The notion of going to the frontiers of space for scientific discovery, business purposes, or military motives requires constant input from creative thinkers, innovators, and competitive risk-takers. Private companies have demonstrated they are more able to accommodate this approach than large government bureaucracies. Thankfully, NASA and other national space agencies realized that private sector partnerships have a great deal to offer, like independent innovation.

In a study by Bryce Space and Technology (2020), only 22.9% ($82.5 billion) of space funding now comes from government sources. The balance (77.1%) comes from private investments in commercial space activities. A full 61% of space investments are non-U.S. This means that American dominance of the commercial space arena has been eclipsed by foreign companies and countries and this share of market continues to accelerate with more offshore space ventures coming online annually. Almost two-thirds (65%) of current space investors are classified as non-space companies (a non-space company is an enterprise whose core business activities are not related to space, e.g., Best Foods, Carnival Cruise Lines, PepsiCo, Caterpillar, Deloitte Consulting, etc.). Most significantly, many firms that are not normally associated with space are actively looking to create new business units to serve the new space sector. Some of these crossover firms see the potential of space as a new market for their modified heavy equipment or for other commercial applications. Other non-space companies are looking to the service sector to provide a broad range of solutions including traditional management consulting, and others are exploring how to serve a potential consumer space market or other vertical business sectors. The Space Foundation's second quarter (2020) Space Report, a guide to global space activity, showed 16.1% annual growth of space infrastructure, 43% annual growth of commercial spacecraft, and an increase of 73% of space activity over the past decade. This growth is especially significant because this robust growth in the space sector occurred in the heart of the 2020 covid-19 global pandemic when most industries were suffering dramatic shortfalls.

Industrial Revolution(s): Space Is the Next Iteration

During the last few hundred years, just a blip on the human history meter, the creation of new technology and innovation has delivered a series of improvements to the world of work. The succession of industrial revolutions resulting from these incremental developments has lessened the reliance on human labor and has replaced the belief that work is a critical social value with a promise of a better, healthier, and longer life through consumerism and convenience. Some of the promises of a brave new world have come true. Some have not.

The agricultural revolution, also known as the Neolithic revolution, started the whole shebang. It began approximately 7500 years ago and changed the way groups of people interacted for the common good by planting, harvesting, and sharing the output of their mutual labor instead of relying on luck as hunter-gatherers. Trade of agriculture products helped to create commercial communities which added trade goods from other market centers. While this development created a new, more permanent, notion of settlement, it also made commercial hubs and whole communities vulnerable to famine from crop failure and targets of war because they were centers of wealth and power. Improvements in agriculture were limited to employing domesticated animals where applicable and the addition of slaves to the labor force to increase production efficiencies. Little changed for millennia. If it rained to make the crops grow, if the locusts didn't ruin the crop, if enemy states didn't raid the grain bins, if the king shared enough of the crop with the peasants to keep them alive, and if the merchant traders showed up to buy the grain at the end of the season, then all was well and good. Why should anything change?

The term industrial revolution is a reference to a major shift in the production of goods or services from one model of work to a more efficient model that reduced dependence on human labor by increasing the application of mechanized means of production. In the case of the first industrial revolution it was the shift from reliance on hand tools and human labor in factories and farms to a more efficient mechanized means of production that was often powered by steam. The positive impact of mechanization allowed people to change their relationship with society and ultimately gave rise to women's suffrage, slave emancipation, environmental awareness, concerns about distribution of wealth, and many other social issues that never had a voice in old feudal systems.

Successive industrial revolutions have transformed the global economy by continuing to apply technology to production and decrease use of human labor. The result has changed social, political, economic, environmental, and ethical relationships. One net result is the increasing dependence on technology, automation, and artificial intelligence (AI). The impact of robotics on the development of space will assign humans to a secondary support role as space infrastructure will be developed and built by robotic automation instead of by human labor. (See Appendix A for a summary of my recent research findings on this and other space topics.)

Subsequent waves of industrialization like the application of electric power for consumer and industrial use, the introduction of the automobile, commercial flight, atomic energy generation, telecommunications, and the internet all built their success because of the successful ventures that come before. The adoption of successive phases of economic maturity in space will rely upon the prior success and achievements of those enterprises that came before.

References

Bryce Space and Technology (Christensen, C., Starzyk, J., Potter, S., & Boensch, N). (2020). *Start-up space; update on investment in commercial space ventures.*

Causey, J. (2019, November 20). *Articles by James Causey|Space.* Retrieved from www.space.com, https://www.space.com/author/james-causey

Ross, W., & O'Connell, K. (2018). *Strategic objective: Expand commercial space activities.* U.S. Department of Commerce.

Part III

Heavenly Markets: Getting Extraterrestrial

This section presents a survey of the economic opportunities in the maturing new space economy. Topics include the role of robotics and artificial intellegence, Earth industries likely to be most successful off-world, and a closer look at specific industrial sectors like mining, manufacturing, agriculture, and others. Also discussed is an overview of space infrastructure issues and the likely challenges of building a business in space.

15

The Industrialization of Space

Why Is the SPCE IPO a Big Deal?

On the morning of October 27, 2019, Sir Richard Branson took his pet project, Virgin Galactic (VG), to Wall Street and put the new space economy squarely on the map. Prof. Matthew Weinzierl of Harvard's B-School commented, "the IPO signals that the [space] sector intends to be more than a niche play … the Virgin Galactic IPO will be seen as proof that the sector is ready for the public markets." Virgin Galactic was listed as SPCE on Dow Jones.

Shelli Brunswick, COO of the Space Foundation, spoke about the importance of Virgin Galactic's historic IPO: "Space is a critical infrastructure with a heavy impact on economic development, entrepreneurial opportunities, and jobs. Wall Street's validation of Virgin Galactic puts a spotlight on space commerce that we haven't seen since the Sputnik era."

Investments in space have rocketed recently as more and more countries, companies, and private investors anticipate huge financial returns from the new space sector. Most of the investment funding has come from the Billionaire's club—Branson, Bezos, Musk, Bigelow—sometimes referred to as philanthropic funding because there is no guarantee the funding would continue if the individual were to leave the stage. Private equity funds have attracted a host of investors to Venture Capital firms (VCs) like SpaceFund and Space Angels. Rick N. Tumlinson, Founding Partner and Chair of Space Fund, comments, "… the time to invest in a company like VG … that is well funded and well managed, is now—before they fly next year."

Most IPO's go through a lengthy process (aka, "the Road Show") where investment bankers like Goldman Sacks make the rounds to institutional investors to sell large blocks of the new stock offering. This series of face-to-face meetings and multiple re-meetings can take a long time to raise the required capital and set the IPO opening stock price.

The new twist is that Virgin Galactic went directly to Social Capital, a private capital investment fund founded by CEO Chamath Palihapitiya, to raise all the money they needed in just 3 months. When interviewed by CNBC reporter Morgan Brennan, Mr. Palihapitiya said the secret was that Virgin Galactic was different because it was, "… in the business of hardware but was a company that performed like a software company." (CNBC interview, October 28, 2019). That's the essence of the new space economy; tech-based operations but competing by creating IP value for stakeholders.

It turns out the new space economy isn't like the old industrial economy after all. Outwardly, Virgin Galactic is the inaugural provider of space tourism. More significantly, this IPO puts space enterprise squarely in the mainstream of commercial investments. Up next, Branson plans to leverage the technology developed for his space planes for deployment to hypersonic point-to-point transportation (think: a trip from the US to Japan in under 90 minutes) with more space-related ventures to follow.

Bank of America's Mary Ann Bartels commented about other space investment vehicles that are showing growth for investors who want to be part of space, most notably ETFs. She noted in a CNBC interview (October 28, 2019) that space-centric exchange traded funds (ETFs) like ROKT, ITA, UFO, and other space-related investment vehicles have shown growth of about 27% in 2019 which is uncommonly strong. While it is important to note that these investment instruments often include traditional aerospace firms like Northrop Grumman, Boeing, Raytheon, and Lockheed Martin, they also represent a strong slice of the current commercial space firms in the burgeoning space industry.

Shelli Brunswick, COO of the Space Foundation, summed up VG's SPCE IPO by saying, "This milestone will hasten the commercialization and advancement of space … for the betterment of all global citizens." And it will open the door for other space ventures to go to the public capital markets and create new investment opportunities in the growing commercial space sector.

16

Space Biz

Space is no longer the domain of American enterprise. Nor is space investment limited to government funding. Bryce Consulting (2020) drilled down into their research and found that 61% of space investment comes from non-American sources. Significantly, of the companies pursuing a commercial space agenda 65% were not typically seen as related to the space industry. This is noteworthy because it validates the view that commercial interests in space are more distributed across multiple economic sectors than just the aerospace usual suspects. It makes sense that a pharmaceutical firm (e.g., Bayer, Gilead) would look to micro-gravity manufacturing in space for a new revenue source. Or that a cruise ship line (e.g., Royal Caribbean) would see opportunities in space as an obvious high margin brand extension. Or that an international construction and engineering firm (e.g., Bechtel) would see the need for building new infrastructure in space as the next big profit center. Or that Rio Tinto would consider asteroid mining a natural strategic next step to their global mining operations supported by the autonomous mining capabilities in development and use since the 1980s by such firms as John Deer, Komatsu, and Caterpillar.

It also makes sense that when the US space agency, NASA, decelerated it's human space priorities, many other nations recognized the strategic importance of space and moved ahead with plans to establish outposts and bases on the Moon, Mars, and other high-value locations in the solar system. Heading the list of national space agencies are China, India, Russia, Europe, UAE, Japan, and a seemingly endless list of countries large and small seeking partnerships with major players. Six agencies at the top of the list have funded active space-related research initiatives that have produced sophisticated technical capabilities like human space flight (the USA, Europe, Russia, Japan, China, and India). In total, over six dozen countries from Australia to Turkmenistan have established official space agencies in hopes of attracting investments and burnishing their national prestige. Strategic partnerships (e.g., Japan, the USA, ESA, and Israel) have created new opportunities for countries and companies to share research capabilities and financial resources.

Trade and commerce will undoubtedly be the backbone of the new space economy. Trading outposts will be established in deep space locations of the solar system far from the more accessible Earth-Lunar axis. The natural satellites of Mars, Phobos and Deimos, are possible staging platforms that could be used to connect mining activities in the main asteroid belt with more developed industrial markets closer to Earth. Certainly, other moons of other planets are also candidates for outposts and communities in a future archipelago of space settlements. Among the commercial sectors likely to prosper in the earlier stages of the new space economy are engineering and construction, transportation and distribution logistics, agriculture, energy generation, human habitat services, and myriad other ventures—some of which have yet to be created.

There are dozens of private space companies, large and small, who are currently focused on serving the growing space industry. While much of their current efforts focus on short-term gains associated with the present commercial space arena of launch capabilities, payload brokerage, satellite communications, TV and entertainment services, data collection, and data analytics, many of these firms have a strategic second act baked into their business plans. This long-term planning horizon looks to the potential of the emerging settlement economy as economically self-sustaining with little dependence upon government funding sources.

In a recent report SpaceFund.com, a venture capital firm focused on capitalizing promising space startups, defined five main areas of viable growth. These are Transportation (119 companies, mostly rocket launch capabilities), Communication (57 companies focuses on satellites), Human Factors (15 companies generally dedicated to space habitats), Supply Chain (25 companies mostly centered on satellite servicing), and Energy (28 companies exploring acquisition and distribution of energy in space). Because of the highly speculative nature of space businesses SpaceFund has developed a *reality rating* for each of the five investment sectors. According to the firm, the goal of the SpaceFund Reality (SFR) rating is, "designed to provide investors, customers, regulators, media, and the industry itself with a quick guide and assessment of players old and new in the various sub-sectors of the space industry" (Crawford 2020).

Energy Generation

One area that causes keen investor interest is producing energy in space for distribution to either space-based users or to Earth. There is active investor interest in this technology because it would signal a dramatic reduction of dependence on carbon-based fuels thus reducing their effect on the environment and would further create opportunities for national independence from petroleum producers. Energy generation could either come from solar or nuclear facilities based in free space where it could be transmitted by laser or microwave directly to commercial customers. This is no small undertaking. A very large investment would need to be made to create production and transmission infrastructure in space as well as facilities on Earth able to safely receive these volatile energy transmissions.

There is a strong current case for beamed power to Earth locations. Sovereign and corporate customers located in remote areas with limited resources to generate their own power or to purchase power from commercial sources could benefit directly. As could far-flung mobile expeditionary military units that may be positioned at sea or in isolated locations. Distribution of space-generated energy could provide immediate power to an ever-changing strategic location. SpaceFund sees great possibilities for the development of efficient power transmission in the next few decades.

Space-Based Agriculture and Food Production

Space agriculture is a growing opportunity. Earlier notions about likely sources of space nourishment included algae, insect larvae, lab-grown meat, chicken, and even 3D printed food (Smithers 2019). Unfortunately, not much has improved the menu. There is, however, some good news about the nutritional value of crops grown in space using local water and, presumably, local regolith (soil). The concept of space-based agriculture is more than adding acreage to the current capacity of global farm production which is of tremendous economic and ecologic benefit. Producing food in space to feed our planet is a potential solution to our growing population. The World Population Clock (2020) expects the number of people to increase from 7.7 billion to over 10 billion by 2060. Bloomberg research on land use (2018) estimates that the USA alone devotes over one fifth of the continental land mass for crop production or 391.5 million acres. This doesn't include the additional 654 million acres used for livestock pasture and range land. The demand for food is outpacing our ability to feed our planet. Space-based agricultural may offer a much-needed solution.

In addition to supplementing Earth's ability to provide necessary nutrition there are other positive long-term benefits of space-based agri-business. These include:

- Reducing the destructive environmental consequences on Earth caused by industrial farming
- Enabling a shift of over 20% of land utilization on Earth from food production to other less damaging uses
- Reducing grazing and raising animals for food consumption and mitigating the impact of industrial meat production on the environment and on human health
- Reducing the calamitous impact of agri-waste (including insecticides and animal hormones that leach into water supplies) on the land, rivers, environment, and climate
- Balancing the opportunity costs of feeding a crowded planet imposed by depleting limited natural resources with other less consuming uses of those shrinking natural resources
- Mitigating the impact of third world primitive agriculture practices like slash-and-burn
- Trading traditional high-cost and inefficient labor-intensive practices for low-cost and efficient automated processes
- Leveraging the broader benefits of transforming the food production sector by initiating less destructive and more efficient methods
- Transforming archaic agricultural methods into a cost-efficient process that produces affordable food on a massive scale to reduce hunger and increase world health regardless of the economic circumstance of the consumer

Current Earth-bound crop production could be supplemented (or replaced) by other sources of easily produced nutrition. Admittedly, the idea of dining on algae-based products, or snacking on processed bugs smacks of dystopian science fiction when compared to other options like 3D printed food and lab-grown meat that seem far more palatable. Non-western cultures have long ago adopted foods like seaweed, insects, and other non-western cuisine that have gained wider commercial acceptance (think: western parochial ideas about sashimi several decades ago). It is likely that, in time with a heavy side-order of marketing, these and other dietary cuisine items will find their way to our global food chain. The specter of Soylent Green notwithstanding.

There are plenty of resources for food to harvest from space that we already know about. One of the reasons China, the USA, and others are targeting the Moon's polar regions is because abundant water in the form of ice has been detected as a resource that could provide humans with sustenance as well as a convenient source of fuel. Mining water from multiple sources in space could become an industry unto its own. Jupiter's moon, Europa, is thought to be a viable source for salt which is a vital ingredient for human survival. The raw materials needed to construct farms on Mars or positioned in space can be obtained from asteroids, moons, and from other plentiful sources in our solar system. Certainly, there is no shortage of sunlight to facilitate crop growth.

Fred Scharmen, writing about human space settlement in *Slate* (2019), noted that the desire to conveniently combine parklike green spaces for both human recreation and agricultural production was not practical. He noted that plants ideally need, "… 24-hour sunlight, constant high heat, and atmospheres overloaded with humidity and CO_2." He observed that in a closed space setting, "A plant paradise is a human hell, not a relaxing garden." Mimicking the meadows and forests of Earth may be a wonderful way to add a sense of Earth to a far-off enclave of space settlers but using those commons to grow crops would be impractical. Scharmen concluded that in space, "… there are two different … purposes for plants and planted spaces [food production and esthetics], and they're not easily reconciled. In space, everything humans bring or make has to have some specific *reason* to be there."

Back here on Earth most of the land devoted to agriculture, about 41%, is for feeding and raising commercial livestock; roughly double the acreage devoted to marketable crop production. This raises the question of whether we should bring our appetite for real meat with us to space. Putting the logistical challenges aside, the production of commercial beef, chicken, pork, lamb, or other animals for consumption in space will be something likely left behind as we venture to space. Animal products like eggs, cheese, and milk may be produced artificially. The good news is that there are several meat substitutes that have arrived on the market. These plant-based stand-ins have already found their way to retail grocery shelves and even to fast food restaurant chains. It is reasonable to assume that these and other similar products will take the place of raising animals for human consumption in space. Aquaculture, farming fish, may prove to be among the few viable opportunities for producing authentic animal protein in space farms.

Some Good News for Space Settlers

Aleph Farms, an Israeli firm, has already demonstrated edible meat can be "grown" in space on the ISS. The process they used collected cow cells to successfully produce cow muscle tissue (beef) in space using a 3D bioprinter. The company's co-founder, Didier Toubia, said, "We are proving that cultivated meat can be produced anytime, anywhere, in any condition." He observed that, "In space, we don't have 10,000 or 15,000 liters of water available [like on Earth] to produce 1 kg of beef." Because the process is portable the artificial beef can be manufactured closer to the consumers needing it, when they wish to consume it, instead of relying on costly shipment and storage. Toubia concluded that the space-based experiment was a step to, "ensure food security for generations to come, while preserving our natural resources." This innovative food technology is an example of how a groundbreaking development originally designed for use in space may quickly be applied to markets here on Earth.

Of concern is whether the Moon, Mars, or another celestial outpost could support the production of sustainable crops of Earth origin. Evidence that supports this capability was recently demonstrated by researchers at Wageningen University in the Netherlands who grew crops using Mars and Moon soil regolith simulant developed by NASA. Researcher Gruyter (2019) demonstrated it was possible to raise crops on the Moon and on Mars to feed future settlers. Of significant importance was their finding that the experimental crops produced viable seeds that could be used to continue a sustainable agricultural process. They grew ten different crops: garden cress, rocket, tomato, radish, rye, quinoa, spinach, chives, peas, and leek. The only crop to fail was spinach.

One of the long-standing tenets of space exploration has been the idea that we should not contaminate the places we explore in space with Earth-based microbes and other terrestrial biological pollutants. When the astronauts went to the moon great care was taken to ensure they wouldn't leave any material that would alter the alien lunar ecology. One reason is that if we discover alien life forms, we need assurance that what we are discovering isn't the careless remnants of an earlier visit by Earth explorers. This idea has been unchallenged throughout the history of space exploration. Until now.

Researchers at Nova Southeastern University published a contrarian view that encourages seeding planets and other locations in solar system that are likely to support future space settlers with Earthly microbes, bacteria, and germs so that the new frontier will be more accepting of Earth crops and other agriculture when immigrants from Earth finally arrive. "Microbial introduction should not be considered accidental but inevitable," says lead author, Jose Lopez (2019). "We hypothesize the near impossibility of exploring new planets without carrying and/or delivering any microbial travelers." This contrarian view is not embraced by current thinking about space settlements but there is a "germ" of logic in this argument. Earthly biology is complex and very much influenced by the environmental context; it's one thing to replicated life in a sterile lab and quite another to turn it loose in the real world where the influence of an assortment of different stimuli can alter the final outcome. Just think of the difference between wine made from grapes coming from the Napa Valley versus wine from the California central coast. They are both great in their own way. But they are very different. [Note: this is an experiment you can easily replicate at home!]

Researchers have explored whether crops grown in space on the ISS are significantly different from Earth grown produce. These findings are important if large-scale space plantations are to come online. This idea maintains that extensive complexes of automated greenhouses could produce immense acreage of crops to feed populations on Earth as well as support a growing number of space inhabitants. Writing in *Frontiers of Plant Science* (Khodadad 2020), the researchers concluded that their study supported the belief that, "… leafy vegetable crops can produce safe, edible, fresh food to supplement to the astronauts' diet." This gives the green light to pursuing a space-based agricultural initiative that could provide nutrition to large numbers of people in need of a ready source of plant-based nutrition.

When space enthusiasts speak about building a sustainable settlement economy one of the major elements mentioned is ISRU or in situ resource utilization. The driving idea behind ISRU is that space can be built with materials harvested from local space resources. The raw materials for building shelter, for energy, and for developing a viable space-based economy can all be obtained in space instead of shipping them from Earth. ISRU is mostly referenced when discussing building infrastructure and constructing industrial complexes. But the notion of ISRU is critically important when planning to feed people in space. All the core elements for growing crops and manufacturing plant-based food stuffs can be found in space. All that's needed are the seeds from Earth and agri-robots to tend the crops.

With increasing evidence that food grown in space for human consumption is identical to the same crops produced on Earth, the commercial opportunity arises for a budding space-based agriculture industry that could not only feed space people but could also provide nutrition to billions of people on Earth. The market for food produced in space may be enormous given the unpleasant facts that (a) our planet's population is expected to reach over ten billions by 2060, (b) climate change has caused prolonged droughts in prime growing locations thus reducing crop yields, (c) the amount of arable land available for crop production is shrinking due overuse and misuse, (d) in Africa and other locations plagues of locusts and other infestations have obliterated food crops and left whole populations on the brink of starvation, and (e) chemical fertilizer technology has approached the upper limit of producing more crop yield per acre. This convergence of issues has created a potential perfect storm where the demand for food will outstrip the production capacity of our planet.

But in space there is no limit on arable land. There are whole planets to cultivate with endless miles of greenhouses or free-floating hothouse plantations suspended in orbit above the surface able to control and maintain perfect growing conditions. Likewise, there is no shortage of water, no locusts to destroy crops, no virgin rain forests to destroy, and no need to artificially enhance crop yield with genetics or chemicals. In short, space-based agribusiness may be able to increasingly supply Earth-based consumers with the food they need to survive as production on Earth decreases, becomes less efficient, becomes more costly, and is less capable of serving the needs of the increasing populations on our planet.

Signposts Up Ahead

In the early stages of adoption space commercial activities will be mostly dependent on Earth. Earth is where financing comes from, where the customers are, where the nations with political will who desire their portion of space are located. But in the later stages of settlement adoption the degree of dependence upon Earth will decrease. Successful ventures in space will steadily create an economic (and social, political, legal) foundation that will compete with Earth-based controlling interests. Some of the indicators, the bellwethers of this shift away from dependence on Earth to a self-sustaining space-based economy might be:

- **Getting down to earth.** The technical problem of shipping large tonnage of goods produced in space down to Earth spaceports will need to be overcome. This is the down-mass problem. If solved it will open markets to space producers and will create opportunities for more investments in space businesses.

- **Looking out for number one.** If the down-mass problem is not resolved in a reasonable timeframe it is likely that space businesses will grow without catering to the consumer mass markets of Earth. The impact of this will serve to create greater independence from Earth. Lack of reliance on Earth will likely serve to energize the space settlement economy.
- **Attention shoppers.** At some point the space settlement economy will shift from an industrial focus to a consumer focus as more and more people choose to live in space settlements. This will change the complexion of trade and will become a challenge to Earth's consumer producers (unless they join the trend and aim for space markets). It is easy to imagine how a solar system packed with consumers could easily bypass Earth's economic leadership.
- **Size matters.** At some point the gross product of space may be hefty enough to alter the center of gravity (CG) of Earth's global economy. A gross space product (GSP) competitive with any of the top ten national producers will earn space delegates a seat at the table no matter what the shape of the table may be (a board room, political representation, an international supreme court, etc.).
- **Warp speed ahead.** The emergence of technology enabling rapid transport of goods and people to the far-flung corners of the solar system will accelerate the growth of the space settlement economy. The net effect of this game-changing transportation capability will be that major shipping hubs will emerge in the farther reaches of the solar system thus creating new population centers that may rival those similar to commercial centers on Earth. A subtle subset of enhanced transport capabilities all over the solar system is the seemingly endless creation of new frontier economies. This will further fuel the growth of the space sector economy.

References

Bryce Consulting (Christensen, C., Starzyk, J., Potter, S., & Boensch, N). (2020). *Start-up space; Update on investment in commercial space ventures.* Bryce Space and Technology.

Crawford, M. (2020). *Energy transmission.* SpaceFund. Retrieved from https://spacefund.com/energy-transmission/

Gruyter, D. (2019, October 15). *Soil on moon and Mars likely to support crops.* Phys.org. Retrieved from https://phys.org/news/2019-10-soil-moon-mars-crops.html

Khodadad, C. L. M. (2020, March 6). *Microbiological and nutritional analysis of lettuce crops grown on the international space station.* Frontiers in Plant Science.

Retrieved from https://www.mybib.com/#/projects/jVX8ly/citations/new/webpage

Lopez, J. V., Peixoto, R. S., & Rosado, A. S. (2019). Inevitable future: Space colonization beyond earth with microbes first. *FEMS Microbiology Ecology, 95*(10). https://doi.org/10.1093/femsec/fiz127.

Scharmen, F. (2019, June 29). *The mindbending challenges of growing food in space*. Slate Magazine. Retrieved from https://slate.com/technology/2019/06/chen-qiufan-space-leek-response-agriculture.html

Smithers, R. (2019, October 7). First meat grown in space lab 248 miles from earth. *The Guardian*. Retrieved from https://www.theguardian.com/environment/2019/oct/07/wheres-the-beef-248-miles-up-as-first-meat-is-grown-in-a-space-lab

World Population Clock: *7.7 Billion People (2019)—Worldometers*. (2020). Worldometers.info. Retrieved from https://www.worldometers.info/world-population/

17

Space Mining

Global mining operations represented $683 billion USD in 2018 with an average net profit of 10% ($63 billion USD) according to Statistica.com (2019). The demand for minerals, coal, rare-earths, and other critical materials makes the mining industry consistently among the top economic sectors on our planet. Yet, the amount of raw industrial materials estimated to be in the solar system eclipses the full potential of Earth's production by nearly an infinite factor! For this reason, mining asteroids and other celestial bodies for raw materials has long been considered a highly likely early business venture in space. The prospect of collecting rocks containing platinum (or gold, or lead), refining the ore for the mineral content, and then selling the purified output to an industrial customer (or managing the production of products via a vertically integrated chain of activities) is very attractive to potential investors. But mineral like gold and palladium aren't the only valued commodities in space. There's water, too (James 2018).

Dr. Carlos Espejel, Space Resources Utilization Engineer at ispace, a private Japanese company that develops robotic spacecraft technologies, noted that because there are tens of billion metrics tons of water on the Moon, "… there is an outer space race between the USA, Russia, China, and India—they are all going [to the moon] for these H2O deposits." The race is on to establish lunar bases at strategic areas of the moon to harvest water and other available mineral resources according to Mining.com (Stutt 2020).

Ye Peijian, the head of the Chinese lunar exploration program, spoke openly in 2017 about expanding the China Silk Road to space and the political motivation behind China's Communist Party views of space. "If we don't go there now even though we're capable of doing so, then we will be blamed by our descendants. If others go there, then they will take over, and you won't be able to go even if you want to. This is reason enough." Some may argue that this is a defensive move rather than a strategic initiative. Whatever the perspective, China's presence in space is assured (Aluf 2020).

Is there Really Gold in Them Thar Asteroids?

The answer is resoundingly "yes." But is it legal to claim ownership in a celestial body and then profit from its resources? In November 2015 President Obama signed a bill that recognized the right of US citizens to own the resources they obtain from asteroids and encouraged the commercial exploration and utilization of resources from asteroids (2015). The earlier benchmark 1967 *Outer Space Treaty* (Wikipedia 2019a), signed by 103 countries, was an attempt to establish legal limits surrounding ownership of property in space. According to the treaty, claiming ownership of an asteroid or other space body is not legal. However, if your plan isn't to claim ownership but to just extract resources and sell them, then you may have legal footing. After all, America brought 842 pounds of rocks back from the moon, that were widely recognized as US property. This is a fine point of law: an asteroid may be completely destroyed in the mining process, reduced to dust, resulting in nothing left to claim or own. At this point the intent of the treaty to protect space territory would become moot. Still uncertain is the issue of legal permission. A license would give legal standing to pursue a legal mining venture in space but neither the USA nor any other country has a recognized process for issuing licenses for space mining. According to Dr. Joanne Gabrynowicz, Professor Emerita at the University of Mississippi, "If you don't have that license, the investors are taking a big chance." Clive Thompson (2016) observed that the value of minerals mined from space may be dependent upon who has ownership rights.

To circumvent the Outer Space Treaty's restrictions on ownership, the U.S. Senate unanimously passed the *Space Act of 2015*, which permits US citizens and companies to claim "non-living" natural resources obtained in space which includes water and minerals. This opens the door for space ventures to acquire and sell raw materials from planets, moons, asteroids, or other celestial sources. It is unclear how this legislation will hold up in international courts. But the Space Act sends a message to other nations that the assets found in space will be fair game (Fulton 2015).

The lure of riches from space makes space mining an attractive business. A relatively small metallic asteroid with a diameter of just one mile is estimated to contain more than US$20 trillion worth of industrial and precious metals.

A smaller asteroid with a diameter of only 1 km may contain over two billion metric tons of iron-nickel which represents between two to three times the world's annual production of the metal. It is likely the asteroid would also yield a small amount of precious metals as a bonus. No wonder a space mining goldrush by the top global diversified mining companies like BHP Group, Rio Tinto, Vale, and Glencore, among the largest mining companies in the world (2019), have a strong interest in expanding their operations to celestial mining (Mahon 2018).

The potential of providing technical and logistic support has not been lost on companies like Caterpillar Corporation, Komatsu, Toyota, John Deere, and other industrial equipment manufacturers. As reported in *Popular Mechanics* (2019) Caterpillar has explored a partnership with NASA for autonomous mining on the Moon. OffWorld, a manufacturer of industrial equipment designed for space, is already planning how to construct mining and other space-based facilities. Current challenges for firms who are planning to conduct space mining include finding suitable asteroid candidates for mining, transporting equipment and labor to and from the mining location, establishing an extraction operation, refining ore at a competitive cost, and selling customers the final product. Unlike operations on Earth, space mining facilities can be moved from one asteroid location to another as needed. Or, in the case of smaller asteroids, the asteroids can be moved to the mining operation.

One of the mitigating factors is the scarcity of an ore on Earth and the spot market price for the ore. As industry usage depletes existing ore reserves on Earth causing prices to increase, the prospect of seeking alternatives in space becomes more attractive. There are economic limits to the benefits of tapping into seemingly unlimited supplies of mineral in the solar system. Suppose your space mining firm located an immense asteroid made of solid gold (or platinum, or copper)—a celestial body roughly the size of Key West, Florida (4.2 square miles). At first, the prospect of cornering the gold market would be a dream come true. But all that gold (or uranium, or palladium) would flood the market, instantly devaluing the precious material and immediately making the mineral virtually worthless. As a result, your company would go bankrupt, stuck with tons of worthless material and unable to recoup operating costs.

Being the first company or country to corner the market in a vital mineral could alter the geo-political balance of global markets. Just as the OPEC oil cartel long held sway over the world's petroleum energy supply, so it is possible that early movement to locate and mine rare and vital industrial minerals may shift the economic—and political—centers of gravity on Earth. It is conceivable that a new space race, a raw materials race, may emerge where competing countries and companies work to prospect for rare mineral deposits in space and to capture monopolistic control of various vital industrial materials.

There are other valuable commodities in space besides gold or iron. There's also water. Lots of it. It turns out that all the water in Earth's oceans came from space in drips and drabs over the millennia carried by random encounters with asteroids and comets (*Scientific American*, 2019). Water is not only vital for human survival, but it is necessary for energy and food production in space. Harvesting water reserves from deep space, from the polar regions of Mars or the Moon, or from countless other places in space will likely become a necessary and lucrative space enterprise. Locating ready sustainable sources of water is key to choosing settlement locations. The abundance of water at the Moon's southern pole is one of the reasons this site is favored by the Chinese and others planning permanent outposts there. Australia has considered mining the moon for water as part of their commercial space strategy (*Bloomberg*, 2019).

Diving for Dollars

The development of the space economy is often characterized as a golden opportunity to access the unlimited mineral riches of the solar system. This thinking maintains that the industrial value of the resources to be found on asteroids, planets, moons and assorted other space sites is incalculable, and that the financial potential gained from exploiting these resources will energize Earth's long-term continued industrial growth for centuries to come. The common refrain is that the solar system is full of a broad variety of valuable metals and minerals just waiting to be picked up and turned into commercial products for industry and consumers. In this view, the prospect of an unlimited mining bonanza in space is the primary reason to invest in space. But there are a few glitches in this business plan.

Why is it necessary to search for ore-bearing asteroids or rich veins of gold, tungsten, or tin out in space when there are still tons of the stuff right here on Earth? Aside from minerals called rare-earths and precious metals like gold most of the minerals that support industrial manufacturing are still relatively abundant on Earth. There is no guarantee that Earth will provide an endless supply in the far future but there is still lots of raw materials available for industrial customers and consumers near term (Lewis 1998).

17 Space Mining

Large-scale initiatives to harvest newly discovered rich deposits of minerals from the ocean seabed have added to the potential mineral harvest here on Earth. W.S. Hylton writes in the *Atlantic* (2019) that the lure of scooping up boatloads of manganese, copper, titanium, nickel, cobalt, and much-coveted rare-earths has attracted strong investment interest for international public and private ventures. These aggressive commercial initiatives aim to tap the mineral wealth lying on or near the surface of the ocean floor. This sounds like a preferable alternative to notoriously destructive land-based mining techniques like strip-mining and using corrosive chemicals for gold mining operations—both of which have well documented damaging impact on the environment, plant and animal species extinctions, and on regional human health and habitation. But the promise of harvesting mineral riches from the sea is not a perfect solution. It will cause destruction and environmental damage that may have an unmeasured cascade effect on the rest of the ocean's ecosystem and, hence, our planet's fragile health. The seabed as a source of raw materials may compete with the exploitation of space-based minerals, but it will not replace it. In fact, the environmental destruction, human health impact, and species extinctions caused by obtaining mineral deposits on Earth may help promote space as a more attractive alternative (Macmillan 2019).

In space, as on Earth, harvesting mineral riches is an expensive and inherently destructive enterprise. It begins with surveying likely places to excavate and extract mineral ore, after the painstaking effort of collecting the raw ore comes the challenge of refining the material into a usable product, and once that hurdle is crossed the finished product must be shipped to customers in space or on Earth. Each of these steps requires a large infusion of capital to build and maintain specialized infrastructure and customized machinery, and an even larger ability of the investor to absorb the risk of operating in the precarious environment of space. Prospecting for mine sites in the solar system, whether located on asteroids or on another planet or moon, will require robotic probes and sophisticated data analytics. If or when a likely deposit is located and the potential value estimated, then mining and refining equipment must be assembled and transported to the mine site. While most of these operations will be automated, it is likely that some minimal human oversite will be required via teleoperations (discussed in greater detail in the following chapter). Humans are needed to be nearby to intervene in the case of system malfunction or to make fine adjustments in the robotic mining operations. Their proximity will be dictated by the need to reduce the lengthy communications lag time from the central commercial centers located at distant places like the Moon, Mars, or some other central administrative location. While the role of humans in space will be nominal due to the high risk of space operations, there will still be a need for skilled technicians to monitor and manipulate industrial mining equipment in real time to ensure accuracy, efficiency, and safety of the operation.

One of the continuing problems that came along with the benefits of the industrial revolution was pollution and environmental degradation. Some industries are notorious for their impact on the environment such as the lead industry, the refining industry, chemical processing, industrial discharge from heavy industrial manufacturing sites, and the petrochemical industry. Industrial manufacturing covers a broad range of activities that often involves pollutants such as chromium, cyanide, mercury, lead, cadmium, and other toxic agents that take a heavy toll on human health and on the environment. In recent years, a toxic industrial manufacturer could avoid fines and sanctions by moving their operation from a heavily regulated environment in the first world to another country—usually a developing country—where oversite was less rigorous and government officials were more accommodating. But the net effect on the planet's environment was the same; pollution is still pollution; catastrophic health concerns still affect humans regardless of their economic setting. If such industrial polluters were to relocate to space instead of simply decamping to another nation on Earth, then the potential net effect of industrial pollution on the planet's health would be greatly reduced or even eliminated. The problem with this business plan is, of course, that clients and customers are on Earth, not in space, and will remain based on our home planet in the foreseeable future. So, the cost of moving businesses that destroy our planet to space far outweighs the benefits of reducing pollution. For now.

Of ancillary concern is the impact of toxic industrial activities on the new space environment. When this topic arises, there is always a question or two from concerned persons about how the pristine environment of space will be fouled with the corruption of poisonous industrial production in space. Will the unsullied setting of space be violated? Yes. And no.

By Earthly measures the environment will be altered. A mine on an asteroid is likely to destroy all or most of the asteroid. A noxious production facility in the void of deep space will produce waste deemed unhealthy on Earth but of little consequence or impact where no life naturally exists at the distant location in space. Waste removal and destruction will undoubtedly be an entrepreneurial opportunity as elements of the new space frontier infrastructure develop and grow.

It is reported that the asteroid Psyche 16 is almost entirely made of metals, including iron, nickel, and gold. One estimate puts the value of these raw materials at roughly $700 *quintillion* (Mahon 2018). There are an estimated 1.1 and 1.9 million asteroids larger than 1 km (0.6 mile) in diameter, and millions of smaller ones in the Main Asteroid Belt. No wonder there is keen interest in harvesting asteroids for valuable materials.

But what about the impact of creating an unlimited supply of precious metals like gold and others? Take the case of the common commercial metal, aluminum (Wikipedia 2019b). For centuries, aluminum was thought to be so rare that its use was limited, and its price exceeded that of gold. But by 1856 a commercial process was developed to convert bauxite into a malleable commodity for wide use in industry and consumer markets. Today aluminum is no longer valued as a rare resource but as a cheap manufacturing material. If the supply of asteroid gold floods the market, then the price per ounce (or megaton) will quickly drop. In that case maybe you can line your oven broiler or BBQ with solid gold foil instead of aluminum foil when you next cook your roast or rack of ribs.

There is much more to the new space settlement economy than gigantic Megatron robotic munchers pulverizing asteroids to extract raw materials. Space offers the more substantial and economically sustainable future promise of a well-integrated commercial network of suppliers, producers, consumers, and corporate clients. Some of the more successful commercial activities will include industrial manufacturing, mining and refining, logistics and transportation, agricultural production and processing, energy production and distribution, and business services and finance. Settled space will have all this and more because settled space will quickly mature from a collection of rugged frontiers into a network of mature economic communities connected by a common platform of trade and commerce. This will spawn the age of the fully integrated space industrial complex.

References

Aluf, D. (2020, July 31). *China's space silk road reaches Mars and beyond*. Asia Times. Retrieved from https://asiatimes.com/2020/07/chinas-space-silk-road-reaches-mars-and-beyond/

Fulton, D. (2015, November 13). *With "off-planet" mining bill, US congress seeks to privatize outer space*. Common Dreams. Retrieved from http://www.commondreams.org/news/2015/11/13/planet-mining-bill-us-congress-seeks-privatize-outer-space

Hylton, W. S. (2019, December 18). Deep-Sea mining and the race to the bottom of the ocean. *The Atlantic*. Retrieved from https://www.theatlantic.com/magazine/archive/2020/01/20000-feet-under-the-sea/603040/

James, T. (2018). *Deep space commodities: Exploration, production and trading*. London: Palgrave Macmillan.

Leman, J. (2019, October 30). Construction company caterpillar wants to mine the Moon. *Popular Mechanics*. Retrieved from https://www.popularmechanics.com/space/moon-mars/a29587959/caterpillar-space-mining/

Lewis, J. S. (1998). *Mining the sky: Untold riches from the asteroids, comets, and planets*. Boston: Addison-Wesley.

Macmillan Publishers Limited, part of Springer Nature. (2019). *Seabed mining is coming—Bringing mineral riches and fears of epic extinctions*. Nature.com. Retrieved from https://www.nature.com/articles/d41586-019-02242-y

Mahon, C. (2018, February 12). *$700 quintillion asteroid ignites space mining gold rush between Mars and Jupiter*. www.outerplaces.com. Retrieved from https://www.outerplaces.com/science/item/17778-700-quintillion-dollar-asteroid-space-mining-gold-rush-mars-jupiter

President Obama Signs Bill Recognizing Asteroid Resource Property Rights into Law. (2015). *Planetary resources*. Retrieved from https://www.planetaryresources.com/2015/11/president-obama-signs-bill-recognizing-asteroid-resource-property-rights-into-law/

Stierwalt, S. (2019, October 6). How did water get on earth? *Scientific American*. Retrieved from https://www.scientificamerican.com/article/how-did-water-get-on-earth/

Stutt, A. (2020, May 22). *The global race to mine outer space*. Mining.com. Retrieved from https://www.mining.com/the-global-race-to-mine-outer-space/

The 40 largest mining companies in the world. (2019, August 10). www.consultancy.org. Retrieved from https://www.consultancy.org/news/145/the-40-largest-mining-companies-in-the-world

Thompson, C. (2016, January 14). *The minerals found in asteroids and faraway planets could be worth trillions. Who gets to mine them?* Wired. Retrieved from http://www.wired.com/2016/01/clive-thompson-11/

Thornhill, J. (2019, April 1). *Bloomberg—are you a robot?* www.bloomberg.com. Retrieved from https://www.bloomberg.com/news/articles/2019-04-01/plan-to-mine-the-moon-gives-australia-opening-in-new-space-era?srnd=hyperdrive

Wikipedia Contributors. (2019a, February 6). *Outer space treaty*. Wikipedia; Wikimedia Foundation. Retrieved from https://en.wikipedia.org/wiki/Outer_Space_Treaty

Wikipedia Contributors. (2019b, June 6). *History of aluminium*. Wikipedia; Wikimedia Foundation. Retrieved from https://en.wikipedia.org/wiki/History_of_aluminium

18

Space Manufacturing

For civilized space to gain economic momentum, to finally stand on its own, several commercial elements need to come together. Considered separately these economic sectors appear vulnerable to independent market pressures. Taken together these threads make a fragile fabric of mutual interdependency. In this integrated model robotics and automation facilitate construction and engineering which supports mining and refining raw materials harvested from asteroids. Brokers and agents find markets for materials and arrange for transportation to producers. Transportation connects the output of asteroid mines to manufacturers in space of industrial machinery, heavy equipment, technology, transport vehicles, and habitats. Humans who are either working in space or living on Earth are nourished by food produced by self-sustaining space farms. Space inhabitants support the various industrial sectors with their technical and managerial skills. In this interconnected value chain are the new opportunities for enterprises that fill the gaps in emerging commercial space economy.

In between raw material extraction and the final consumer is space-based manufacturing. As of this writing it is assumed that most production output will supply the demand of companies and consumers in space because the technical limits of shipping large commercial cargo from space to Earth are currently prohibitive (this is the down-mass problem, as discussed earlier). If the primary customer base is in space, then it is reasonable to assume the complexion of demand will reflect the scope and the types of goods and materials produced in space. If the initial commercial activities in space are primarily industrial, then the products produced to meet space-based demand will also be primarily industrial. Later, as more people settle space manufacturing output will shift to serve the individual consumer.

If technical limits are overcome and Earth markets come to dominate space-based manufacturing, then it is more likely the types of goods produced will supply consumer demand in markets on Earth. This distinction is important. A consumer market is a mass market that often demands competitive pricing, low margins, greater competition, and distribution channels that must accommodate returned and unsold goods. An industrial or B2B market, like the space-centric market previously discussed above, carries different pricing, higher margins, longer term contracts and service agreements as well as customized product modifications.

Jeff Bezos, the creator and owner of Amazon and Blue Origin Federation, LLC (AKA: Blue), has a vision of space that recognizes heavy industrial manufacturing as the top reason to create a human presence in space (Foust 2019). His vision of why it is important to go to space is about helping to heal Earth rather than abandon it. One component of that plan is to reduce the significant degradation of the planet's environment caused because of heavy industrial manufacturing by moving these toxic and hazardous activities to space (Bezos 2017). It is this unique differentiator, in Bezos' vision, that sets space apart from Earth. The fundamental assumption of *Blue* is our world needs time to heal, to set itself right, to lay fallow and rejuvenate. But that doesn't mean we have to shut down the planet, instead we should transfer the environment-killing industrial activities to space where the impact is minimal. He is not alone in this belief.

Some of the new technologies and innovations that original equipment manufacturers (OEMs) are introducing for the factory of the future that will facilitate space manufacturing include ultrasonic machining, 3D virtual-reality imaging, digital scanning, and further advances in additive manufacturing (Warwick, 2017). Because many suppliers in the value chain are ill-prepared to adopt these innovations there will be a reshuffling of how new manufacturing technologies affect their long-term sustainable competitive advantages in the marketplace. These and other manufacturing advances will become the staples of the next wave of space-based manufacturing.

Manufacturing in space isn't only about producing more efficient fiber optics cables, pharmaceuticals, and solar technology. The space environment also offers the advantage of unlimited solar energy, the unique manufacturing environment of micro- or zero-gravity, and the benefit of producing uncontaminated products in a pristine vacuum. This is especially important for 3D bio-manufacturing of health-related products like human hearts, skin, and other replacement organs. Nearly 7600 heart transplants were performed globally in 2017, but there's an immense lack of donated organs to fill this need, causing thousands of people to die every year. The promise of printing a custom-made heart or other malfunctioning organ in space labs could save thousands of lives annually (Woollacott 2019).

Using a giant 3D printer to construct a space freighter, the components of an asteroid mining operation, or a self-contained settlement outpost is a big undertaking. Building an aircraft carrier or a large container ship on Earth takes thousands of workers representing hundreds of specific trades. But in space specialized robots can do most of the work. A few people will oversee the manufacturing production process as "teleoperators" working in front of monitors displaying the schedule progress and auditing quality assurance. In some cases, a "cobot" relationship will pair a human with a highly specialized robot to perform delicate or intricate oversite.

Teleoperation

There is some debate about the role of humans in space. In the new space frontier, there is little need for traditional human labor, no need for farmers tilling furrows, no demand for housepainters, little need for mechanics. People will be needed to support the machines that will run the day-to-day life of the settlement. Robots will operate the mines, fix the other robots, construct whatever needs to be constructed, and generally manage all the labor-intensive responsibilities. Robots can work 168 hours per week instead of the paltry 40 or 70 h that human workers produce. Robots don't need health care. Robots don't need food or extensive vacations. If robots get their periodic maintenance checkups they will work as required. So, what will people do? How will they have the satisfaction of being a productive contributor? Thankfully, there will still be important roles for people (Densford 2017). One of the roles that will be necessary is teleoperation.

Teleoperation is the remote human intervention in controlling an otherwise automated manufacturing process. Operating a mining operation far removed from the home base will increase risks of problems and missteps due to the lengthy time to communicate commands to mining equipment. If a probe is misplaced or a load of ore misaligned to the hopper then production may suffer. To correct for this eventuality, it will be necessary for humans to be near to the machines to guide and override operations, if necessary. This teleoperation process is akin to crane operators managing cargo containers at a modern port of entry. The operation could be automated but it is safer to have an operator manage the process of selecting and moving containers to and from the dock and the ship.

Much of space exploration has already been managed via telerobotic control. The only people involved in the space mission after the launch to space are the folks communicating with the equipment in space from back on Earth. The greater the distance from Earth the longer it takes to communicate with the probe or rover and the longer it takes to interpret pictures and telemetry, the longer it will take to send corrective commands. One suggestion about exploring Mars or other distant locations in the solar system is to park a human crew in orbit above the planet to manage robotic exploration on the surface. This will not only decrease the response time to the vehicle on the surface of the planet but will also ensure increased safety for the crew. The process is much like playing a sophisticated video game. The person controlling the robot machine may wear a virtual reality (VR) headset or simply view a video screen and operate the movements of the equipment with manual controls like flying a drone. MIT developers have created a VR system for controlling remote operations of automated machines by using a commercial Oculus Rift headset.

The idea of tele-robotics has been applied in space and in fields like medicine for some time. When a space probe is sent to Mars or to the edge of the solar system, a human technician back on Earth monitors its progress and sends signals to the spacecraft to alter its trajectory in space or to steer a rover over to inspect an interesting feature on the surface of the planet. Future mining operations will likely be controlled by human intervention in a similar way because the ability for immediate response from a nearby operator will allow for more instant control when faced with a technical problem or an unexpected opportunity. Telemedicine, a specific form of teleoperations, is gaining broad acceptance, especially since the covid-19 pandemic curtailed face-to-face medical visits. Remote consultations and surgical operations are also gaining broader acceptance, with the surgeon controlling an invasive procedure from another room or even from another country.

An example of a product manufactured in space that is superior to the same product manufactured on Earth is ZBLAN fiber optic cable. Crystals, microscopic flaws, that occur in optical fiber make transatlantic communications, lasers, high-speed internet, and other applications commercially infeasible. Zero-gravity production was tested on the ISS in early 2018 and showed that the tiny crystals that often appear in fiber made in Earth's gravity were completely absent (Caughill, 2017). ZBLAN, a high-grade virtually flawless fiber optic cable manufactured in space by the Made in Space company, demonstrated how micro-gravity manufacturing can produce superior products for a variety of space and Earth commercial applications.

Humans may not get to play the role of wrench-turners in space, but they will still have positions of final authority over the output of the production process. They will still run enterprises and build new communities. But they won't live a life in space that was anything like their familiar life back on Earth.

Robots and Cobots

A fascinating subset of the future space workplace is the integration of robotic workers with humans. The mix of humans along with robots, or *cobots*, opens the door to a new landscape of managerial challenges. If you are a manager with both robotic and human workers how should you evaluate individual productivity. More thought-provoking is the potential challenge of navigating the potential relationship you may have with your robotic supervisor. Do you buy your cobot boss a bottle of wine for the holidays or a vintage can of WD-40? More to the point, will such professional relationships be merely transactional, or will they escalate to a level of transformative interactions?

Answering Question #10

Then there is the unknown territory of workplace peer-to-peer relationships. Gallup, Inc., a management consulting firm, famously conducted a study about workplace retention that determined, unsurprisingly, the number one reason that employees *leave* a job is because of conflict with their boss. But the surprising finding was that the primary reason for *staying* at a difficult job was question #10 on the questionnaire. This question asked about the relationships established with coworkers that extended outside of the workplace. If your new best friend at work is a cobot, an automated coworker, what will be the likelihood of bonding with it and establishing a meaningful relationship? A *Wall Street Journal* video documentary on the future of workplace automation highlighted an MIT study of manufacturing efficiency using a combination of human and robot workers. The study looked at three production teams: Robot only, human only, and a mix of robots and humans working together on the same production line (cobots). The surprising finding was that the most efficient production team, after accounting for technical variances, was the combination of both humans and robots. This was partially attributed to the dramatic reduction of human worker idle time when teamed with robotic coworkers. It seems humans tend to assert their competence through competition (Hernandez, 2018).

As cobots become more integrated into our personal and professional lives how will people answer question #10 if their close friend is a cobot? For that matter how will the cobot answer question #10?

References

Bezos: In Future, Heavy Manufacturing Will Take Place in Space|Aviation Week Network. (2017, March 3). Aviationweek.com. Retrieved from http://aviationweek.com/space/bezos-future-heavy-manufacturing-will-take-place-space?eid=forward

Caughill, P. (2017, April 12). Zero-G space-based factories may be the future of our world. *Futurism*. Retrieved from https://futurism.com/neoscope/zero-g-space-based-factories-may-be-the-future-of-our-world

Densford, F. (2017, October 12). *MIT researchers develop VR teleoperation robot control system*. The Robot Report. Retrieved from https://www.therobotreport.com/mit-researchers-develop-vr-teleoperation-robot-control-system/

Foust, J. (2019, March 5). *The cosmic vision of Jeff Bezos*. SpaceNews.com. Retrieved from https://spacenews.com/the-cosmic-vision-of-jeff-bezos/

Hernandez, B. (2018, September 24). *The robot revolution: The new age of manufacturing*. ETF Trends. Retrieved from https://www.etftrends.com/robotics-ai-channel/robot-revolution-new-age-manufacturing/

Warwick, G. (2017). *Technologies for Factory of the Future | aviation week network*. Aviationweek.com. Retrieved from https://aviationweek.com/aerospace/connected-aerospace/technologies-factory-future?Issue=AW-020_20170125_AW-020_736&NL=AW-020&cl=article_3&elq2=9a45df3b71b347d69dab8a46e0f2d6e9&sfvc4enews=42

Woollacott, E. (2019, January 22). *Why your new heart could be made in space one day*. BBC News. Retrieved from https://www.bbc.com/news/business-46944972

19

No Country for Earth Men

In case you're thinking about ordering your custom-tailored bespoke spacesuit from Amazon for your new life in space, don't hold your breath because the odds are against you or any other human working in space in the foreseeable future. That's not to say there won't be lots of work going on in space, just not so much work done by people. There will be mining, manufacturing, energy production, communications, and even vast agricultural plantations to feed humanity. There'll be lots going on in space. But there just won't be lots of people doing the work in space. Sophisticated AI-driven robotics will be the likely space workforce for decades to come, but not you. Sorry.

Make Earth Great Again

The political humorist Bill Maher made a case for people NOT going to live on Mars on his HBO program, *Real Time* (2017) when he showed his "*Official Scientific Chart*" explaining the benefits of staying put on Earth. He made a simple but eloquently contrarian point that going to Mars was a dumb idea. Unlike Earth, Mars didn't have breathable air, readily available food, potable water, protection from ambient radiation, sophisticated medical resources in case of emergency, or all the other things that make life on Earth worthwhile. Unlike Mars, Earth had air, food, water, and all the esthetic beauty of sunsets, rain showers, meadows green with grass, and the sounds of birds to punctuate the day. While not everyone agrees with Maher's skeptical ideas about space settlement there is something truthful in his observation worth noting. Why build a home in the wilderness of space when you already have a place to live here on Earth?

It turns out space is downright deadly—jam packed with OSHA hazards. So, unless you're getting that spacesuit to go to space as a weekend tourist or a refugee fleeing from yet another global pandemic the prospect of working or living in space is pretty darn slim. But why is the notion of space exploration so connected to the idea of people doing all the work and all the exploring in space? For over a century Sci-Fi books and films have influenced how we think about our future in space. Many fantastical fictional concepts have turned into technologic realities; like reusable rockets, moon landings, and rovers doing all that roving around on Mars. It only seems natural that if we can do all the whiz-bang technical space-age stuff then people are guaranteed to be part of the action. But just because make-believe spacemen have adventures in space doesn't mean it's going to happen soon for the rest of us; space just isn't a people-friendly place.

O'Neill's Legacy: Expectations of Eden

Physicist Gerard K. O'Neill's writings about man's future in space influenced a generation of future space scientists and engineers with his ideas about human space settlement. Unlike the legions of Sci-Fi writers who came before, O'Neill had the pedigree of a real PhD and the ability to posit a plausible vision of man's future in space. His very popular and richly illustrated book, *The High Frontier: Human Colonies in Space* (1976) combined fictional descriptions of people living in space with his very real plan to build space colonies where adventurous colonists would "live and work in space." Talking with Baby Boomer space scientists who came of age when his book was most popular, I often heard bits and snatches of the High Frontier vision that still hold sway today and has influenced everything from hardware design to space policy. O'Neill's vision included graphic illustrations of immense space wheels where whole communities live, work, grow food, and do whatever one does when living out in space. O'Neill is still a revered reference of senior engineers at NASA and other space-related organizations. In O'Neill's future space habitats there are trees and shrubbery growing in the filtered sunlight, there is a meandering endless stream that winds for miles around the inside of the transparent wheel held in place by the artificial gravity created by the wheel's slow rotation, and there are space settlers jogging alongside the unending stream bed and doing whatever space colonists do when trapped in a beautiful celestial prison for years at a time. It is a vision of space utopia. But it's not going to happen soon. Dr. O'Neill may have painted a graphic vision of futuristic space that inspired and seduced a generation of space enthusiasts, but it will be at least another generation or two until we will be ready to live in O'Neill's Eden.

The High Cost of People in Space

From an economic perspective it's cheaper to operate a space business (or any business) without the expensive upkeep and headaches that come with supporting lots of human employees. People are a notoriously high maintenance expense and make the whole space exploration thing messy and complicated. Significantly, when humans go along for the ride the top priority of the mission shifts away from accomplishing the goals of the project to keeping the human crew alive at all costs. Without the people component, the priority of the mission would be to simply get the job done. A manned expedition to Mars won't be about merely getting to another world in our solar system, the mission's top priority will be to bring the crew safely back alive. The infrastructure investment for sustaining human life makes the break-even calculation prohibitive.

Prioritizing human life wasn't always the norm. In prior centuries, when ships set sail to traverse the planet it was a given that the boat would come back with fewer shipmates than when they started: some would fall overboard (or get pushed), others would escape to live with friendly folk on a tropical isle, a few would "walk the plank" for misbehavior, and others would just drop dead from scurvy, malnutrition, or some other fatal disease. But when things got rough the captain didn't call a town hall meeting to take a vote about whether to continue. He stayed the course.

Humans currently work somewhere between 40 and 60 productive hours per week (not counting morning coffee breaks, bathroom breaks, lunch breaks, internet browsing breaks, planning meetings, status report meetings, afternoon coffee breaks, sick days, vacation days, national holidays, time on the internet searching for another job, time spent complaining to HR about how unreasonable the boss is, and personal mental health days). People also need the company to kick in for costly health insurance, vacation time, unemployment insurance, workers compensation, matching 401 k contributions, and so forth. On the other hand, a robotic device can be programmed to work nonstop for all 168 h in a week with an occasional pause for maintenance and an invigorating lube. People at work have all sorts of competitive issues with one another, want periodic raises and bonuses, constantly complain that they aren't getting the work/life balance they desire, and generally make life miserable for their managers (who often make life miserable for *their* boss). Robots don't need managers, only programmers and a visit to the maintenance bay every 6 months to upgrade their software and tighten a few loose screws. People tend to change jobs thus incurring the high cost of talent acquisition to replace, retain, train, and to keep good talent. Robots don't take time off to "find themselves" and, unlike humans, can be conveniently disassembled and neatly recycled for parts at the end of a project.

The good news is that most of what we want to do commercially in space can be done via automation. Manned space enterprise may be compared to the challenge of navigating the LA freeways going to and from work every day. The freeway obstacle course is more about surviving the traffic ordeal than getting to the office to accomplish stuff. Companies have found that when they take the commute out of the daily grind, employees can focus on productive work-related tasks instead of spending all that time listening to NPR and calculating the days until retirement. Likewise, unmanned missions easily avoid the entire drama of human survival and just focus on doing the tasks at hand.

Without all the life support and communications gear required to sustain people a space mission is leaner, simpler to execute, and cheaper to implement. Because high risks earn high rewards unmanned space activities can take much greater risks and therefore potentially deliver much higher rewards for investors. Robots are disposable and expendable after all—just another expense line, a capital asset that can be depreciated, on the mission budget. Still, space exploration by remote control without human heroes just doesn't fit with our O'Neill-driven image of what we expect about life in space. We assume that space exploration is a human enterprise. We don't assume that space exploration means robots boldly going and doing exciting stuff, where no robot has gone before.

For all the financial and practical benefits of robots, it's very unlikely that people will be removed from the process of settling space even though it may be economical and efficient. It is possible to take people out of the space exploration equation for the time being—but it is not likely that settlements will fully exist without people. The image of space settlement is inexorably connected with the commonly accepted notion of people bravely carving a new civilization out of the space wilderness and the belief that, regardless of risk and added cost, it will be people who will play the starring role in developing space settlements. Putting this noble vision aside, it certainly would simplify things and reduce the start-up costs if the foundational work were to be accomplished by robots to clear a path so that people can come later. The prospect of space as a literal "no-man's land" poses an obvious question: If the new space economy favors robots and automation over people, then who—or what—will ultimately be in control of man's destiny in space?

Making space civilized will require a tremendous effort. Just a short century ago, building vital infrastructure like railways and subways, sewer systems and water distribution systems, railway cars and motor cars, tilling the soil, and building skyscrapers required armies of men to do the heavy lifting, metal bending, bolt twisting, and every other aspect of the industrial enterprise. Not so in space.

Building space will be a job for industrial technologies and robots. Self-managed robots and autonomous equipment will take the place of men and women from the front line of construction to the back line of logistics and quality assurance. This isn't to ensure human safety, although that is a beneficial advantage, but to expedite the complex challenge of completing large-scale projects as economically efficient and effective as possible. People get tired and lose focus, drop tools where they shouldn't be dropped, fail to follow the schematic diagrams in detail, and tend to implement their own ideas instead of following the plan. Robots do whatever their programming dictates and don't show up late or intoxicated for their work shift.

The advantage of using machines instead of people is widely employed for mining operations and factory work here on Earth. The use of robots and other automation in space has been anticipated by industrial equipment companies like Caterpillar, Komatsu, Hitachi, OffWorld, Volvo, and Sandvik for decades (Woods 2019). They see opportunities for semi-autonomous graders, loaders, bulldozers to build roads, buildings, infrastructure, and landing sites. Excavation equipment can operate without human control to mine water, regolith, and other materials used to manufacture building materials at the job sites. All the critical components are already in place to operate robotic construction including AI, LIDAR, GPS, and other onboard diagnostic capabilities.

As discussed previously, cobots, specialized robots designed to work collaboratively alongside human coworkers, will be used to alleviate the monotony of repetitive tasks and to manage more dangerous tasks on the production line. Cobots are designed to assume six primary manufacturing tasks: packing and packaging, tending a programmed machine process, assembly of parts into a subassembly or finished product, palatizing finished goods for shipment, operating in a clean room environment, and serving as a work assistant to a human worker or team of workers. The cobot is an adjunct to either humans or to other robotic workers. Human/Robot interaction (HRI) is an area of study that explores how humans enter relationships with an automated colleague, or cobot. Current deployment of cobots occur in elder care, medical/therapeutic settings, industrial production, autonomous driving, and an increasing range of other social and manufacturing uses. Space exploration offers opportunities for HRI on long space voyages (Think of the HAL 9000 computer in the 1968 film, *2001: A Space Odyssey*, but without the homicidal tendencies), for interaction with autonomous rovers and construction equipment, and communications with unmanned drones providing a constant stream of feedback about potential points of interest just over the horizon. Use of cobots in military operations has gained greater interest and research continues to explore the depth of HRI relationship and dependency, especially in times of stress.

A recent example of how cobots were used to supplement critical professional labor (e.g., physicians and other medical personnel) in short supply and to dispense vital services to patients occurred in early 2020 during the covid-19 pandemic in China. Specialized cobots were used to conduct diagnosis on a mass scale without the assistance of human medical personnel (Hornyak 2020). Robots and other internet of things devices (IoT) performed all necessary medical screening services. Each new patient who entered the specially designed facility was screened and evaluated based upon temperature data and other biometric vital signs like blood pressure, heart rate, and blood oxygen level. The medical cobots were connected to the facility network so that humans (i.e., doctors and nurses) could access individual patient information to determine current and progressive state of health and risk to other patients currently undergoing care in a facility and in the greater community.

All basic diagnostic medical services in the facility were carried out by the cobots. Patients entering the facility were screened by 5G-connected thermometers to alert staff for anyone with a fever. Patients wore smart bracelets and rings that synced with CloudMinds' AI platform so their vital signs, including temperature, heart rate and blood oxygen levels, could be monitored. Doctors and nurses also wore the devices to catch any early signs of infection. But not all cobots were created equally. Other specialized robots dispensed disinfectant and cleaned floors. Another set of cobots distributed information, food, drinks, and even medicine depending on individual patient requirements—all without the intervention of the professional health care provider.

References

Hornyak, T. (2020, March 18). *What America can learn from China's use of robots and telemedicine to combat the coronavirus.* CNBC. Retrieved from https://www.cnbc.com/2020/03/18/how-china-is-using-robots-and-telemedicine-to-combat-the-coronavirus.html

Maher, B. (2017). New rule: Make earth great again|real time with Bill Maher (HBO) [YouTube Video]. In *YouTube.* Retrieved from https://www.youtube.com/watch?v=mrGFEW2Hb2g

O'Neill, G. K. (1976, 2000). *The high frontier: human colonies in space.* Apogee. (Original work published 1977).

Wikipedia Contributors. (2019, March 25). *Human–robot interaction.* Wikipedia; Wikimedia Foundation. Retrieved from https://en.wikipedia.org/wiki/Human%E2%80%93robot_interaction

Woods, B. (2019, October 23). *Caterpillar's autonomous vehicles may be used by NASA to mine the moon and build a lunar base.* CNBC. Retrieved from https://www.cnbc.com/2019/10/23/caterpillar-and-nasa-developing-autonomous-vehicles-to-mine-the-moon.html

20

Colonies, Outposts, Settlements, and Stations

Over past decades there has been much research and thinking about what a space outpost or settlement would look like, the kinds of activities that would take place there, and the types of people who would live there. Pulitzer Prize recipient Gerard K. O'Neill brought much of this into focus when he wrote *The High Frontier* (1976). His book sparked a serious discussion about the technical feasibility of living in space and inspired many to think of space settlement as something realistic, attainable, instead of just a topic of science fiction. His most enduring legacy is his technical justification for building giant cislunar (space-based not planet-based) habitats that not only housed its citizenry but also acted as a greenhouse for production of food and parklike green spaces. His proposed large wheel and cylindrical habitats in space set the tone for what has now become modern space settlement study.

There are different approaches to building a space habitat. Some, as advocated by O'Neill, could be built in free space and thus claim no planet's real estate. Other models would be constructed on the surface of a planet like Mars, on a moon like our own or on or in orbit around one of the 176 known moons in our solar system, or even on or adjacent to an asteroid of which there are thousands. In a simplified overview, no matter the location of the community there are common considerations that include infrastructure (transportation, utilities, landing sites, and other shared support services), manufacturing production facilities, human shelters and habitats (personal dwellings, commercial and governmental offices, etc.), life support (food production and water purification), and an economic platform for sustaining the settlement community (trade and commerce).

An ideal habitat would mitigate the dangers of living in a non-Earth setting and would have to manage the extreme conditions of space including life-threatening lethal temperatures, rampant ambient radiation, micro-meteorite damage, and the challenges of acquiring food and water. Because of the general hostile setting in space robots would play a significant role in the early phases of habitat construction. Automation will play a key role in space productivity well after the settlement habitat is constructed. Plans for construction would rely upon in situ resource utilization (ISRU) which assumes the use of materials found mostly at that site or proximate (Sibille 2012). Planning for ISRU construction removes the prohibitive costs of transporting building materials from Earth or from another far-off outpost. The assumption is that viable construction materials could be fabricated using industrial-size 3D printers.

An alternative to heavy assembly using construction materials whether shipped from Earth or obtained ISRU are inflatable modules, which have already been tested extensively on the International Space Station (Emspak 2016). The Bigelow Expandable Activity Module (BEAM) is a privately manufactured inflatable room that can be quickly set up for temporary or even permanent use in mid-space or on the surface of a planet or moon. BEAM modules also provide required protection from radiation and from micro-meteorite hits. A larger module is the BA-330 habitat that accommodates 330 cubic meters, or 11,653 cubic feet, of usable volume inside. Bigelow's BA-330 habitat is designed to protect occupants from space radiation and space debris. The habitat unfurls into a two-story, 55-foot-long (16-m-long) outpost that could house up to six astronauts. BEAM's builder is Las Vegas-based company, Bigelow Aerospace Corporation (Howell 2016). Four other companies have also built inflatable prototypes and are competing for the NASA contract to provide Gateway with human habitat modules. They are Boeing Co, Northrop Grumman, Sierra Nevada Corporation, and Lockheed Martin. Each of these firms received a portion of the $65 million set aside by NASA to jumpstart development of the inflatable prototypes (Roulette 2019). This is a relatively small portion of the space agency's proposed $500 million budgeted to kickstart development of the Gateway project. Each competitor has integrated unique user-friendly features like windows, bathrooms, an exercise area, a small kitchen, and noise-canceling sleep stations. Not to be outdone, the Bigelow inflatable habitat boasts two toilets.

Cislunar City

One of the great things about outer space is that there is a lot of space out there. More empty space than all the planet surfaces combined. Just as the concept of self-sustaining floating cities captures the imagination of creative futurists looking to expand humanity's footprint here on Earth, the idea of cislunar settlements marries some of the best ideas about creating new space-based habitats with efficiency. Strictly speaking, the term "cislunar" refers to the space between Earth and the Moon, but the term is sometimes loosely used when referring to interspace outposts almost anywhere in space where the structure isn't connected to the surface of a celestial body. Sticking to the literal definition, a cislunar outpost in orbit around the Moon would conveniently serve as an intermediary gateway connection point between Earth, the Moon, Mars, and the rest of the solar system (Vedda 2018).

NASA plans to build just such cislunar gateway (creatively named The Gateway), or Lunar Orbital Platform, to manage mission logistics to Mars, to lunar surface installations, and as a platform for voyages to deeper space destinations. Gateway would be more than a stop-over. In time it would likely become a central transportation logistics hub serving the trans-shipment of materials, manufactured goods and people. Gateway could evolve into a complex community of services like a full-service freeway rest stop connecting critical destinations on the space frontier and providing vital services and technical support. This smaller version of the International Space Station orbiting the Moon is likely to grow and expand into a mini city in space over time.

United Launch Alliance (ULA), a private joint venture of Lockheed Martin Space Systems and Boeing has proposed an alternate cislunar model, the *Econosphere*. Although ULA believes it will take decades to fully realize its full vision, the long-term plan is to invite a broad range of commercial partners to participate in a self-sustaining community of around 1000 people in the cislunar space between the Earth and the Moon. The goal of the privately funded venture is to become self-sufficient and economically sustainable, like an industrial park in space where tenants generate rent and share a common infrastructure. It appears that the cislunar economy will be the incubator for a fully formed cosmos economy, and stations in space, around the Moon or Mars or beyond, are the first stage of a multi-stage process of economic development in the cosmos economy (Bergin 2018).

Finding the Zone

As of 2019 over 4000 exoplanets had been identified (an exoplanet is a planet in another solar system beyond our own solar system). Before 1992 there was no concrete evidence that planets existed outside of our solar system but since then with the help of Kepler and the James Webb Telescopes it's become apparent that our galaxy, our universe, is ripe with all sorts of planets (Moyer 2013). Some of them are strong candidates for Earth-like habitation.

The hope is to find planets in the "Goldilocks Zone," a planet that's not too cold, not too hot, not too small, not too big, with an average temperature that's just right and supports liquid water on its surface. Even if there are no animate alien life forms to meet, a planet with these characteristics could support life from our planet; in other words, us. As we cast about for extraterrestrial places to explore and settle, the best option would be a surrogate Earth-like planet where we could easily adapt without having to rebuild the living conditions to suit our delicate biology. But until we find such a place, we'll have to consider other options, some of which are in our own backyard.

We have the makings for creating our own cislunar goldilocks communities. If we look to the Earth's orbit as an optimal zone for sustaining human life (not too hot, not too cold, etc.), then building cislunar cities that are strung out in the approximate region of the Earth's orbit may provide a network of habitable gateways to the rest of the solar system for exploration and development. These cislunar islands could be specialized around industrial sectors (e.g., agriculture production, large-scale industrial manufacturing) or could be future urban centers housing off-planet populations and a broad spectrum of commercial enterprises. In time, some of the larger cislunar cities may develop their own network of satellite "moons" of private communities or corporate enclaves.

Space Immigrants

It's hard to accurately describe the types of persons who will migrate to space but it's a lot easier to understand why they want to go. Getting a clearer picture of who will sign up for a one-way trip to space is a matter of matching the different visions of space settlement with the different profiles and motivations of likely space immigrants. Some will be motivated to construct a lifeboat in space for a floundering planet stuck in an ecological downward spiral. These pilgrims will see going to space as an act of conscious altruism. Others will see the current impact of pandemics, famine, war, global warming, and other catastrophic events as a sign that life on Earth is no longer tenable; that it is time to abandon ship. For them going to space is a matter of species preservation and survival, a way to stay one jump ahead of certain disaster. Still others, will go to space for the adventure and the challenge of being among the first to establish a beachhead for humanity in a new frontier. Their agenda is purely personal, a matter of fame and recognition.

The initial wave of space settlers may probably come from backgrounds well-suited for Earth's white-collar jobs; educated idealists with a strong sense of purpose about the rightness of what they are doing, a vague grasp of the risks they are taking, and a willingness to abandon comfortable lives and relationships they will leave behind. Some of these intrepid first wavers will find space to be what they were looking for, a place of endless opportunities and challenges that will provide a sense of purpose. Others will find the harshness and deprivation of a life on the edge too much to accept, much more than they bargained for. In due time a following wave of settlers will come as hired hands employed by companies or enlisted by nations to carry out the difficult task of laying the groundwork infrastructure for industry and keeping order on the space frontier. They will come with technical expertise and training that will depend upon an ability to work closely with automation and robotic equipment. This group may not come to space with a desire to permanently settle but, as contractors, may decide to stay in space when their time of service expires. The third wave of space immigrants will come to space seeking a new life and commercial prospect. They will probably come seeking a chance for better opportunities and the hope of building wealth in the emerging space economy. This third wave of settlers will open the frontiers and build the communities for their families and for future space residents.

Each of these waves of immigrant migrations to space will display its own unique type of courage befitting the collective personality of the cohort. The more rugged and less mature the space habitat, the more individual raw courage will be required. Any trek to a new life in an unknown land requires a mix of blind faith that all the risks taken are worthwhile and that regardless of hardships suffered everything will work out well, that the privations will be worth the reward, and that luck and fate will deliver you from harm's threats. But a pilgrim who migrates to space in the first wave where nothing habitable is yet built nor any sense of community is established, will require a much different type of courage—a courage that is more of a commitment to a purpose or spiritual calling than just a joy ride to a new adventure. The ability to believe in making something new, as an act of altruism, is far different than the later phase courage that comes of leaving a familiar comfortable life on Earth for an unfamiliar life in the wilderness of frontier space.

There is something exciting and risky about following a dream. There is something exhilarating about taking a step forward into your destiny with no looking back. Like falling in love or finding your calling in life. People have always uprooted themselves to seek something better beyond the next hill. Better hunting, better climate, a better life. Those courageous enough to take the journey to a new life were either rewarded with success or destroyed. At the heart of their quest was risk and courage.

Courage is not the act of denying fear. It is acknowledging fear and making fear your friend. With an embrace of fear comes an intimate knowledge of your own limits and who you can become. The space immigrants traveling to the outposts of the solar system will not live far from their fear, but they will live closer to who they are and wish to be.

A Settlement by any Other Name …

Figure 20.1 describes the comparative positioning of the major space community formats in terms of relative autonomy and economic sustainability.

The names for places in space where people and/or robots gather to do work, meet to conduct financial transaction, or just congregate for the sublime social benefit of being around other sentient beings may go by various labels. A place in space could be called a colony, an outpost, a station, or a settlement. Some may be fixed on a planet, on a moon or in a predictable cislunar location and others may be mobile in free space. Whatever the location there are similarities and there are differentiating distinctions. This is my effort to clarify these distinctions because now, on the cusp of the new space economy, these labels will become important.

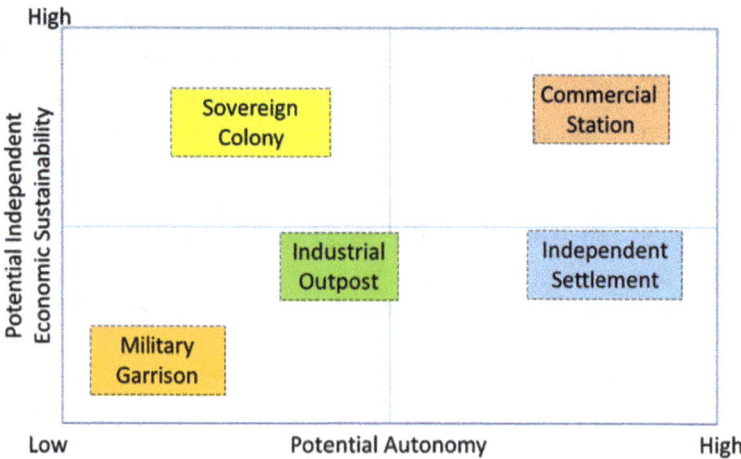

Fig. 20.1 Space Communities. A comparison of several space community formats measured by independent economic sustainability and the degree of relative autonomy

Colony

The term used most often to describe a future space community is *colony*. It is the default general term most often used when describing any form of human settlement in space. I use the term colony here to represent a community or set of communities under direct economic control and political influence by an established nation. I borrow this historical notion from the model of colonial expansion employed in recent centuries when sovereign states like Denmark, Spain, Britain, Germany, France, and other countries sought exclusive access to natural resources, manufactured goods, and agricultural products, beyond their traditional borders. Crown colony control was often enforced by military force. After independence, social and political norms such as the legal system, economy, language, and social rules were often adopted by the former colony. The colony of British Guyana (now Guyana) mirrored British law, used British currency and banking rules, spoke English, and modeled Britain as the prototypical system of government when they achieved their independence. Next door, the colony of Dutch Guyana (now Suriname), took its standards from the Netherlands, just as another neighbor, French Guiana known more for its infamous Devil's Island prison, took its governance cues from France and now hosts the Guiana Space Centre, the European Space Agency's (ESA) primary launch site near the equator.

Colonies typically followed a mercantilist model of exclusively trading with the governing nation as a form of extended protectionism. For example, the global network of colonies in the British Commonwealth tended to conduct trade primarily with other British Commonwealth colonies (hence the term indicating a shared or "common wealth"). Settler colonialism, so named because this form encouraged the in-migration of citizen-settlers from the governing nation, may be one template likely to develop in space. Also likely is surrogate colonialism, where the colonists are recruited from other nations and locations, not necessarily from the controlling colonial power. For example, Canada, Australia, and other former British colonies openly accept new citizens from other nations.

Outpost

Unlike a colony, an outpost may be viewed as a less formal, less structured location in space. It is formally defined as an outlying or frontier settlement, a remote branch or position of a main organization or group. A colony, as we've defined it for our purposes, is owned or controlled by a sovereign nation. An outpost, defined in the context of space, is likely to be a functioning economic entity owned and controlled by a corporation. Outposts traditionally come in military and/or industrial flavors. Trading outposts on the fringes of civilization in frontier Canada and America often served doubly to establish a commercial and a military presence.

Station

A station is a central hub of commercial and social activity where specific tasks take place. This is very similar to an outpost except that a station is more generally accessible to a wide range of people from different companies, nations, and settlements where an outpost, as we define its use in space, is restricted to military or closed corporate use.

Settlement

A settlement, by comparison may be defined as a place where people come together to establish a *community*. The defining characteristic of a settlement is its civic cohesion, its sense of collective identity, where the resident population considers the settlement to be their home, where children are educated, where a robust self-sufficient social structure defines the settlement's sense of identity. It is a social group of any size whose members reside in a specific locality, conduct shared government, engage in a local economy, and often have a common cultural and historical heritage. In this way a kibbutz-like model may be an example of a settlement because of its sense of collective ownership, shared governance, and idea of success for the entire collective community.

Colony, outpost, station, and settlement are terms that are often loosely used interchangeably when referring to centralized places in space where people gather to live or work. They may be out in free space or anchored to a planet, moon, or asteroid. A cislunar location, for example, refers specifically to, the space between the earth and the orbit of the moon, but, on occasion, refers to a station location in deep space either in orbit around a body or positioned farther out in the solar system.

Given time, luck, access to resources and individual ambitious leadership, it is possible for a deep space station, an industrial outpost, or a colony to transform itself into an independent, self-determining settlement. Such is the history of the past several centuries in the developing New World as colonies changed hands from one nation to another and then ultimately achieved independence.

Space communities will remain relatively small affairs for some time. Large metropolitan centers may not fully develop until there is enough critical mass of trade, access to capital, talented workforce, and infrastructure to support a large population. In early stages, these small communities will be fundamentally self-sufficient relying on the types of resources locally available and the entrepreneurial capital investments it attracts. The closer the community is positioned near a major economic hub, the greater its interdependence with the flow of goods and services from that hub. If the community is isolated (by distance, inability to offer competitive goods for sale, contentious political affiliation, etc.) then the community will probably retreat to greater autonomy in the larger space economy. In this case it will be, by necessity, more spartan in its use of limited resources and energy, thus further inhibiting its ability to grow and survive.

It is probable that community affiliations will blur these neat and tidy definitions. Rival states and territories have sometimes come together in détente as a cooperative federated format in places like the United States and The European Union, for example.

A noteworthy difference between a space community and a community on Earth will be the abandonment of individuality. Home ownership, already on steady decline in the USA in favor of renting, will be rare in space, along with the acquisition of durable consumer goods for individual use. The socialist model of the early Israeli Kibbutz where all material goods were communally owned may be a likely template in space for early communities. Consider the impact of the collective social structure on creating a business model aimed at that target market. If the goal is to appeal to a collective decision consensus (instead of establishing a targeted plan to reach an individual consumer), then the product design, its warrantees, marketing message, and support services will need to be structured accordingly. On Earth an individual in a condo or an apartment may easily purchase kitchen and laundry appliances for personal use and replacement. But for a communal setting like some space settlements, the requirements may be for a more heavy-duty product of less esthetic appeal, a product having more durable design characteristics in order to provide reliable utility that will serve multiple users over a longer term. The marketing strategy required for those buyers will need to account for multiple inputs on the buying decision instead of just appealing to a single decision-maker as is the case on Earth. The product design and sales transaction process will be much different for a group purchase than if the end user were an individual.

Of note, also, is the choice of best business model. In the case of serving a collective community customer in a remote location it may be mutually beneficial to lease or rent goods rather than purchase to own. A variant of this is a "rent-to-own" option where payments contribute to the purchase of the item. In this model less capital is needed up front, but the final financial outlay may be far greater over an extended contract than a direct purchase because of added fees and interest payments.

Space Peddlers

Providing goods and services to outlying outposts and settlements may prove to be a successful business model on the space frontier. In the nineteenth century, new American immigrants tramped the wilds of the midwestern frontier to sell a broad inventory of much-needed dry goods, consumer durables, remedies, eyeglasses, drugs, hardware, furniture, and nearly anything else a farm household could desire when unable to easily travel to a local village (Linard 2004). The convenience provided by these back-packing dry goods merchants served a necessary need on the frontier. Some of the more successful peddlers earned enough to send for the rest of their family who were still living in a foreign land. Some started a general store in town and establish roots in their new homeland. A notable few founded large mail order and department store chains (e.g., Sears, Montgomery Ward, Goldwater, and others) as the frontier matured and became more civilized (Diner 2014). Some form of space peddling may likewise fill the need to bring home goods and services to isolated space pioneers.

Affiliation and Identity

There is no guarantee that the governance of space settlements will be either benevolent or benign. Advanced technology will empower opportunistic leaders with the ability to control social order with an absolute hand. Likewise, there is no guarantee at all that space settlement leaders will adopt a democratic approach or manage with the values of social justice in mind. Given the potentially powerful role of private firms in space it is probable that space settlements could end up functioning like old-fashioned company towns where every aspect of human life in the local economy is dictated by prescribed policies. Workers who sign contracts to work in the far reaches of the solar system will be cut off from the safety and convenience of legal recourse and redress. They will likely be considered a captured asset or ad hoc indentured labor.

References

Bergin, C. (2018, March 22). *ULA laying the foundations for an Econosphere in CisLunar space—NASASpaceFlight.com*. NASASpaceFlight.com. Retrieved from https://www.nasaspaceflight.com/2018/03/ula-laying-foundations-econosphere-cislunar-space/

Diner, H. (2014, May 28). *German Jews and Peddling in America*. Immigrant entrepreneurship; German Historical Institute. Retrieved from https://www.immigrantentrepreneurship.org/entry.php?rec=191

Emspak, J. (2016, November 22). *Blow it up: Inflatable space station habitat shows promise in early tests*. NASA. Retrieved from https://www.space.com/34798-bigelow-expandable-habitat-space-station-test.html?utm_source=sp-newsletter&utm_medium=email&utm_campaign=20161125-sdc%20and%20 https://www.space.com/19311-bigelow-aerospace.html

Howell, E. (2016, February 26). *Bigelow aerospace: Inflatable modules for ISS*. Space.com. Retrieved from https://www.space.com/19311-bigelow-aerospace.html

Linard, L. (2004, April 19). *Birth of the American salesman. HBS working knowledge*. Retrieved from https://hbswk.hbs.edu/item/birth-of-the-american-salesman

Moyer, E. (2013, May 23). *Requiem for Kepler? NASA's pioneering planet-finder (pictures)*. CNET. Retrieved from https://www.cnet.com/pictures/requiem-for-kepler-nasas-pioneering-planet-finder-pictures/2/

Roulette, J. (2019, October 17). *NASA eyeing inflatable space lodges for moon, Mars and beyond*. Reuters. Retrieved from https://www.reuters.com/article/us-space-exploration-habitat-idUSKBN1WW1H2

Sibille, L. (2012, July 3). *ISRU home*. isru.nasa.gov. Retrieved from https://isru.nasa.gov/

Vedda, J. (2018). *Image courtesy of NASA cislunar development: What to build-and why center for space policy and strategy*. Retrieved from https://aerospace.org/sites/default/files/2018-05/CislunarDevelopment.pdf

21

Don't Look Back

Is there room for human compassion in the rugged space frontier? Consider why an immigrant immigrates. Either life is abhorrent or dangerous where they live so they flee their old home, or the promise of a new life in a new place offers a better life so they take a risk, abandon all that's familiar, and travel to their new home. The motivation for change is to trade a miserable life for the *opportunity* of a better life. With no guarantees that the new life in the new place will be any better. In fact, the common thread of immigrants over the years is that they are making the sacrifice for future generations of their family and not for themselves.

There are other motivations for space settlement. Classic economics long held that people will always act in their own best interest (ethical egoism). But this once iron-clad maxim was challenged by Nobel Prize winner Elinor Ostrom in *Governing the Commons* (1990) and by others who cite unselfish acts of altruism, behaviors where a person may put themselves in harm's way in order to help save a stranger from injury or death, as examples that refute the notion of ethical egoism.

The prospect of a tight-knit settlement community made up of individuals acting solely in their own interest would be untenable and evokes a reference to *The Tragedy of the Commons* metaphor (W.F. Lloyd, 1833) where the shared resources (the commons) would be quickly depleted if each member of the society acted without concern for the others in the group. Recall those who refused to wear a mask during the 2020 Covid19 pandemic and consequently put the lives of others at risk. These individuals rejected the notion of the commons (society in general) as a place of mutual benefit where individual precautions would benefit the larger community.

Economic opportunity isn't an end, it is the means to an end. It's not really about the promise of wealth, but the hope that with the acquisition of new wealth will come a better life, a safer life, a happier life, for themselves and for others.

In 1919 a Russian immigrant named Morris Kurtzon had achieved some success with his electric light manufacturing business in his new American homeland. He had come from poverty and now wanted to provide a service for others that would last as a legacy in his Chicago community. Consequently, he founded Mt. Sinai Hospital in what was then a poverty-ridden Jewish community on the west side of the city so that the indigent members of the community could receive the health care that was not available to them because of their inability to pay or because of religious prejudice denied them access. Today Mt. Sinai Hospital serves a Latino and African American mostly indigent community and subsidizes 97% of their medical costs. Mo Kurtzon's legacy is a demonstration of leveraging social need with economic opportunity to benefit future generations of strangers. The lure of riches in the new frontier of space is more than a matter of money. The immigrants who take the leap to space will create homes and communities that will outlive them and that will serve the others to follow. There are characteristics that will likely set space immigrants apart. Some will believe their move to space will be part of a movement to heal an infirmed home planet. Others will go because they may believe that space is humankind's manifest destiny or any number of other possible visions that serve their bold act of leaving Earth for a new life in space. But whatever the reason for leaving, the rewards will probably be paid to those who will follow; it is the future generations who will benefit most.

Benefits Back on Earth

The development of space settlements will enable a shift off planet of several heavy industrial sectors, thus reducing their negative impact on earth. New innovative technologies developed for the space environment will be adopted by industrial users on Earth in order to stay competitive. Industries supporting the establishment and growth of industrial space will likely become an intellectual property (IP) incubator for many of the traditional industries based on Earth, thus raising the level of productivity for both Earth and space economies. For example, Mitsubishi and Caterpillar, both makers of large-scale industrial construction vehicles, have adopted technologies they've already developed for space-based construction to their Earth-based equipment designs, thus measurably improving production capabilities. Modern large-scale agriculture grain production has a tremendous negative impact on local ecology and on the Earth's climate. Space-based agri-business operations will provide an alternative opportunity to mitigate the pollution caused by large-scale agricultural production by moving these resource-heavy operations to off-planet space farms. This will reduce the impact of industrial farming on Earth's air quality, water pollution, and emission of greenhouse gassed into the atmosphere.

There is a growing trend in business over the past decade to adopt the practice of implementing the "triple bottom line" as an ethical corporate value. This approach adds the very real costs of social responsibility to a company's financial statements to take ownership for the impact on the environment as a corporate citizen. The traditional bottom line is simply a number that shows the difference between revenue and the expenses incurred to produce a product or service. This is translated as either profit or loss. The triple bottom line adds additional expenses like the estimated impact of the company's production on the environment and the costs associated with social impact. These three costs are sometimes referred to as people, planet, and profit. Naturally, these additional line items influence the bottom-line number either by decreasing net profit or by increasing net loss. Companies that adopt this metric feel they are making a positive value-based statement about their relationship to the environment and to society.

As commercial space settlement takes hold there is likely to be greater acceptance of this socially conscience practice. Especially by those entrepreneurs and investors who consider their efforts in space as a chance to create a better path forward for humanity and a chance to correct past exploitation of the home planet. Caught in the crosshairs of entrepreneurial altruism is the question of what this ethical challenge may say about the promise of space settlement vs. the reality of making and sustaining business success (i.e., profit).

Let's not forget that the essence of space settlement—the essential element of the proposition—is about people. At the core it is a humanist movement. Uprooting a comfortable life on Earth and beginning a new life in a harsh frontier is an act of faith that the new life will be better somehow, that the act of settling in space is a broad political statement as much as it is a deeply personal statement, that even though it is highly probable that a life in space won't be better than the prior life on Earth the process of settlement will create something lasting: a legacy of something bigger than the self. This is the nagging risk that Morris Kurtzon took when he left for America. There were no guarantees he would survive let alone prosper. But he evaluated the risk and calculated that the benefits outweighed the costs for him and his family.

From Globalism to Space-ism

The vision of networked space settlements throughout our solar system can be misleading. Space will not likely become settled in a way compatible with our notion of civilization today. Many settlements will be frontier outposts similar to trading posts of earlier American adventures, others will be corporate installations where only employees, vendors, and customers will be welcome, still others will likely be extensions of political/military establishments (e.g., a Chinese base with their own garrison), and some, expectantly, will be independent communities; a crossroads of trade and migration.

Because of settlement diversity, the concept of universal law and a common definition of justice will be fluid. In addition to the risks of living in space and the risks of growing a business in a fragile economy, there will likely be the risks that come from dishonest and dangerous actors operating where there is little chance of protection or legal sanction. This risky insecurity will change in time as the phases of adoption move from a frontier economy to the later stages of adoption where more infrastructure, competition, and financial permanency will create a more stable social culture.

Much of our current thinking about space references today's sovereign states playing a leading role in the exploration, exploitation, and colonial settlement of space. This is a holdover from the infancy of the space race between Soviet Russia and the United States when government pride and rivalry unleashed a torrent of funding to be the first to put men on the moon among other milestone accomplishments. But, more recently, the private space industry sector has recognized the high profit potential of space as a new source of revenue. This anticipation of the new "globalism 2.0" has seen the dramatic shift from state sponsored space activities to private companies dominating commercial space.

Fully 61% of all global space investors are non-US based companies and 65% of space investors are non-space firms. This means that companies instead of sovereign governments are dominating the growing space business sector and that private companies investing in space are chiefly made up of enterprises not normally associated with space, aerospace, or defense. Space investment is now attracting a range of firms associated with space tourism, off-world construction and engineering, pharmaceuticals, and a variety of other companies whose brands thus far have not been associated with space-related business such as Rio Tinto Mining, Mitsubishi, and even Princess Cruise Lines.

It is one thing to make a pledge to your boss (to cut costs in your department, to meet a sales target, to launch a new product by a specific date, to show up for work on time every now and then), but it is another thing to make a commitment to yourself. Both avenues of behavior focus on accountability. Both are contracts to perform at a high standard in pursuit of an objective that serves a measurable outcome. But the commitment to yourself has greater impact because it is non-negotiable and speaks to your core values. Hitting your sales target will look good on your annual review with your boss but making time in your life to pursue a graduate degree or spend more time with your family supports personal life goals and values of your choosing and has a lasting effect. Thus, it will be for those who commit to space. It is something done to resolve a personal desire; a value that speaks to what they hold important in life, a promise fulfilled.

A Shift to the Cosmos

The development of space settlements will usher a new epoch in human history. The growing influence of off-planet populations will be more than a new marker on human history's timeline; it will signal a major shift in our home planet importance for the members of our species. Space habitation means that we will no longer be fully dependent upon neither the limitations, nor the benefits, of planet Earth. As human populations in space increase and move further from our historic home it is likely that individuals will adopt to their new environment and identify more as a citizen of their settlement than with their place of origin on Earth. This potential shift in social identity will represent the coming maturity of the space settlement economy. When immigrants begin to identify with their new home and forgo the customs of their old life, they take on a new perspective, a new individuality, apart from the old order of things and become, for example, hyphenated; Irish-Americans, Italian-Americans, African-Americans, XYZ-Americans. Will settlers on Mars self-identify as Canadian-Martians, Israeli-Martians, or Japanese-Martians? One day, maybe people will see themselves simply as Cosmic-Humans.

Reversal of fortune. At some future point, the critical mass of space settlement may come to dominate commercial enterprise in the solar system. When this occurs, Earth's role and influence will likely become subordinate to one or more other leading space-based economies. This shift of influence and power, while not unique in our human history, would create a new narrative in the timeline of human history. Just as the fall of Rome left a legacy of politics, art, religion, and social order for subsequent powers to adopt and modify, so too, will the diminution of Earth's economic and political influence open the door to other emergent societies of humans located elsewhere in the cosmos to rebrand the earlier values and customs of the human species.

The end of the Earth Age. Humans have bumped along from Prehistory and the Stone Age, through Ancient History and the Iron Age, to Modern History with the Industrial Age, the so-called Atomic Age, the noteworthy Early Space Age, and the Information Age. These historical slices may now best be summarized and labeled as the Earth Age. But the advent of space settlement across our solar system may well bring about the end of the Earth Age and inaugurate the start of the new Solar Settlement Age.

You can't go home again. The famous title of Thomas Wolfe's novel (1940) has become a catchphrase describing the effect of the passage of time on the way things used to be. You can't go home because home, and the people who live there, have changed, as have you. Nothing stays static. You can't go back so don't look back.

For settlement pioneers there will come moments of buyer's remorse. Regret for a life on Earth forfeited for the adventure of space. Some will adopt to the harsh reality of living in primitive conditions. Others will go broke trying to build a business where products and customers don't fit. Still others will simply give up the dream that used to inspire them. That caused them to risk everything for a life on the new settlement frontier. But, the trip to space will be a one-way ticket.

When destitute immigrants came to America in the nineteenth century from Italy, Japan, Russia, China, Ireland, and other nations, it was a one-way trip. They knew that there was nothing for them to return to in the old country; when they walked off the boat they were committed to a life in the new world. A hard, solid line was drawn between their old life and their new home. That will likely be the same for those who choose a life as a space settler.

Reference

Ostrom, E. (1990). *Governing the commons: The evolution of institutions for collective action*. Cambridge: Cambridge University Press.

Part IV

The Rise of Power, Politics, and Policy in Space

The new race for profit and power in the solar system will pit nations and companies against one another for dominance and survival. New alliances will form, former partners may become adversaries, supply chains will develop to serve a variety of space sectors. A discussion of the S-Curve model and how it represents the maturity cycle of the space economy sets the stage for an overview discussion of how the space economy will be adopted by producers and consumers. The section closes with a discussion of the impact of opening the space frontier and space settlements and how space-based enterprises will influence traditional commercial activities on Earth.

22

The New 49ers Rush to Space

At the Tipping Point

At the dawn of the electric age in the late 1880s, Thomas Edison and Nikola Tesla were commercial rivals advocating opposing notions about the best method of distributing electricity to an eager customer base. Edison advocated the use of more costly but much safer direct current (DC) while Tesla was a proponent of cheaper but less safe alternating current (AC). The resulting legal battles lasted well into the twentieth century (AC prevailed). But the high cost of the rivalry and legal battles that resulted slowed the adoption of a standard vision for national electrification (Sandford 2012). Ultimately, the world got the benefits of electricity, but both entrepreneurs lost control of their inventions and depleted their cash reserves in endless legal squabbles. One wonders what would have transpired if the two rivals decided to collaborate and find a solution that would satisfy their sense of ego and the needs of the customers they served. The potential downside of competing visions can be self-destructive, or it can yield creativity and new opportunities.

There are many visions of what space may likely become, optimistic visions of settlement and financial growth, apocalyptic narratives about the end of our civilization. The range of expectations reveals a lack of a unified game plan for commercial space settlement; no central strategy has emerged that drives collaboration across national or corporate platforms. Without a unifying master plan that consolidates the good intentions and best thinking of the major private companies and countries, each will be left to themselves to determine their own best course of action, an exercise in self-interest. They may share a mutual interest to develop and settle space, but they have yet to collaborate to develop a shared roadmap they all can follow to space.

Competition is a double-edged sword. Without competitors a firm may control a market sector. With competitors the firm will be challenged to produce better, cheaper, and more effective products. The costs of the political competition between Russia and the United States in the mid-twentieth century (i.e., the Cold War, the Space Race) sparked massive government investments in military technologies and seriously depleted cash reserves. But the benefits found their way into non-military commercial applications that helped fuel the economy. Today's internet (originally named Arpanet which stands for Advanced Research Projects Agency Network) grew from a military effort to reduce the vulnerability of critical government communications. As a result, our internet-based global economy grew out of Cold War competition for military security and supremacy (Wikipedia 2019a).

A benefit of commercial competition is the rich diversity of ideas, resources, and talent that private sector companies and sovereign countries can bring to the process. Competitive spirit is a good thing from a free market perspective—it fuels the fires of imagination and adds a sense of urgency and importance to the project. But competition can also create inefficient waste and redundancies. The Space Race of the mid to late twentieth century saw billions of dollars (and rubles) spent to develop very similar military capabilities. The high opportunity costs of assigning specialized engineering and scientific talent to military applications severely constrained Russia's economy. These and other vital resources may have been better deployed in other beneficial areas. Instead, the political culture of waging a win/lose competitive rivalry curtailed economic growth and crippled social programs in both societies (Wikipedia 2019b).

The two nations' only shared vision was the scorched-earth policy of winning the Cold War by making the other nation lose at all costs. This take-no-prisoners approach was embodied in the goal of "winning" the space race by betting on the success of Apollo 11. Putting an American on the Moon was a goal for its own sake. Because the Moon landing was a short-term achievement and not a stepping-stone for future manned space exploration, no one has visited the Moon since December 1972 (Loff 2000). After the last astronauts returned from the moon there was no plan (or political will) for humanity to go further into the solar system. The *race* itself was the goal. Getting men on the Moon ahead of the rival Russians was more important than conquering man's dream of space exploration.

If the true job of leadership is to create a vision of success and then gather the resources to accomplish that vision, then space settlement will not happen unless there are leaders who are willing to work to achieve the vision of human settlement in our solar system. Fortunately, today there are leaders from nations and private companies who share a passion for human space exploration and settlement. Leaders who are working to make their vision a reality. Unfortunately, like Edison and Tesla, they are often reluctant to join forces and share resources to craft a common narrative, a decisive roadmap, to efficiently achieve that vision together in common cause.

What might have come from Tesla and Edison joining forces? What might yet come from Musk and Bezos pairing their efforts, or from China and the United States partnering together with a common plan for space settlement?

Globalization 2050

Space is no longer the exclusive domain of the U.S. or any single nation. Space activities around the globe have mushroomed and spilled over borders far beyond the economic or political control of just a few nations. Government funding of space ventures has been dramatically eclipsed by private investors. Bryce Space and Technology (2020) reports that over 77% of today's space expenditures come from *non*governmental sources. Large and small countries who have not previously developed space programs or space-friendly financial incentives have increasingly taken note of the potential economic gain from promoting space investments. These smaller sovereign states may not have the technical capabilities or research potential of China or the U.S., but they do possess a desire to not be left behind in the new space economy and a willingness to entice space start-ups to their shores to promote economic development.

Some of the more noteworthy recent international space partnerships illustrate how large and small countries and private companies are finding ways to participate and profit in space via collaboration. Some recent headlines illustrating the international diversity of space activities are as follows:

- Joint **Japan/India** moon mission. The Indian Space Research Organization (ISRO) and Japan's Aerospace Exploration Agency (JAXA) are partnering for a joint mission to the Moon. This is the second time Japan and India have combined efforts for space exploration.
- **United Arab Emirates** (UAE) Mars mission. The UAE used a **Japanese** (JAXA) rocket to launch their mission. They partnered with the **University of Colorado** at Boulder, **Arizona State University**, and the **University of California at Berkeley** to develop the science instruments. The Emirates Mars Mission Hope probe is designed to gather data for future Martian

exploration in the early 2020s. The mission was successfully launched in July 2020.
- **European Space Agency** (ESA) members including the **French** space agency (CNES) and **Germany's** agency (DLR) along with **Japan's** JAXA are planning to test their reusable launcher from 2021 in Kourou, **French Guiana**.
- Over 80 **Chinese** commercial start-ups have begun competitive operations indicating that the emerging Chinese space sector is expanding at a rapid pace and may soon be able to compete with Western launch and payload providers.
- **China** plans to employ robotic technologies to construct a moon base by 2025. The next step is to have inhabitants living in the facility by 2030, thus establishing mankind's first permanent lunar outpost. (Unless someone else gets there first.)
- The **UAE** plans **to establish a self-sustaining habitable settlement on Mars by 2117.**
- **India** plans to launch its first manned mission to space in December 2021: a three-astronaut, 7-day flight. India has arranged collaborative agreements with **Russian** space agency (Roscosmos) and **French** space agency (CNES).
- **Japan** (JAXA) is leveraging their time on the International Space Station (ISS) to enable growth of commercial space activities. Japan is home to over 40 space start-ups; three companies raised more than $100 million each. Hiroshi Sasaki, director of JAXA Space Exploration Center, says, "We want to establish key collaborations for missions. JAXA cannot do everything by itself."
- **U.S.** and **Japan** are negotiating to send Japanese astronauts to moon in 2020s.

At least 72 out of 195 nations have created a national space agency. (Yes, even Luxemburg has a very active space agency!) Admittedly, some of these agencies are figureheads for national pride and prestige. But many have been developed with an eye to economic development and the hope of attracting foreign partners and investors from both the public and private sectors. Some new space nations see participating in the space economy as a must-do mandatory investment to support their long-term national economic sustainability.

As the commercial space sector grows and expands across the globe, more specialized business plans have been launched. In addition to launch and satellite production support services like data analytics, mission control, payload aggregation, and space traffic control have come to market. Launch sites closer to the equator in places like French Guiana offer some economic advantage by taking advantage of the centrifugal movement of the planet. Many of the firms targeting current customers have their sights set on the future of space commerce beyond Low Earth Orbit (LEO). As the sector expands and gains momentum so, too, will the types of business ventures that will seek rents and revenue from space.

References

Bryce (Christensen, C., Starzyk, J., Potter, S., & Boensch, N.) (2020). *Start-up space; Update on investment in commercial space ventures*. Bryce Space and Technology.
Loff, S. (2000). *Apollo 11 mission overview*. NASA. Retrieved from https://www.nasa.gov/mission_pages/apollo/missions/apollo11.html
Sandford, M. R. (2012, July 10). *AC/DC: The Tesla–Edison Feud*. Retrieved from www.mentalfloss.com. https://www.mentalfloss.com/article/30140/acdc-tesla%E2%80%93edison-feud
Wikipedia Contributors. (2019a). *ARPANET*. Wikipedia; Wikimedia Foundation. Retrieved from https://en.wikipedia.org/wiki/ARPANET
Wikipedia Contributors. (2019b, January 18). *Space race*. Wikipedia; Wikimedia Foundation. Retrieved from https://en.wikipedia.org/wiki/Space_Race

23

A Competitive Solar System

Trade in Space

There is little doubt that eventually the solar system will become a vibrant economic engine. Wealth from mining, manufacturing, and trade will grow along with increasing numbers of outposts and settlements. Local economies of space communities will capitalize on their production capabilities and on the transient shipment of raw materials and finished goods. This is not unlike our Earth economy. There has been serious thought about the impact of great distances in our solar system on such elements as the calculation of interest charges, insurance, and other financial elements that are time dependent. Nobel Prize recipient Dr. Paul Krugman published a paper, The Theory of Interstellar Trade (2010), in which he cautioned that because very high velocities of data transmission near the speed of light will alter time, financial transaction calculations would have to be adjusted accordingly, *"because the time taken in transit will appear less to an observer traveling with the goods [being purchased] than to a stationary observer."* We are light-years away from encountering this problem.

The Role of Competition

When similar problems are approached by people who envision different solutions amazing things can happen. Using the earlier example of Tesla and Edison who were famously at odds with each other over which design would dominate the transmission of electricity, each man pursued a different solution to the same problem. Along the way new hurdles were encountered which required new creative answers. In chasing their individual vision of how to achieve the same goal each innovator created new patents and developed technology solutions that enriched the understanding of how to efficiently generate and manage electricity. It could be argued that if it weren't for their intense rivalry, their desire to win at all costs, the discipline of electrical engineering might have taken much longer to develop and mature. But competition can be both constructive and destructive. Sometimes it boils down to a matter of judgment.

There are healthy benefits to operating a business in a competitive market. First, the best competitors can motivate a firm to stop being complacent and get out of their comfort zone. A market leader may lose sight of innovation and be surpassed by a new entrant if they fail to see the changing economic landscape. Second, a competitor may force a company to be more innovative to meet changing customer demand. Third, the competitive environment may energize a company to be more aggressive seeking new products and new customers. Fourth is the benefit that comes from the diversity of thinking that emerges from the pursuit of solutions to new problems that have not been encountered before. Different people see the same problem with their own perspective and apply their unique creativity and logic. That's why we have different car designs, different house architectures, and different styles of clothing. That is why bringing about a unified vision for man's future in space, a plan that appeals to different people with different ideas of space settlement, requires more than a summary of all the technical engineering requirements. Finally, competition creates opportunities for partnerships and M&A (mergers and acquisitions). At some point the best options are to merge, acquire, or be acquired by another firm. This consolidation may salvage long-term customer relationships but also may destroy the core of your organizational culture. None of these benefits assure success but each holds an opportunity for improvement and a chance to be a better competitor (Mburugu n.d.).

Just as there are decided gains from operating in a competitive market, there are also advantages establishing a noncompetitive, or monopolistic, position. First, because a monopoly is positioned to make an excessive profit margin, excess capital can be deployed to maintain the monopoly position or be diverted to other areas of the operation such research and development. Second, monopolies have a reputation for charging excessive prices just because the customer has no alternative. Think of the Duty-Free shops in an airport that charge extremely high prices for items that are readily available for much cheaper prices at a big box store like Costco. Third, monopolies have the luxury of not worrying excessively about internal cost controls. Because they are the only game in town there is no pressure to introduce more efficient production technologies. But this is a fool's game and can backfire when the inevitable competitor enters the market with lean cost controls, better service, and cheaper pricing (Amir 2015).

There are also disadvantages of a monopoly. Because customers have no alternative and must buy from the monopoly, the customer must suffer the high prices. This usurious practice does nothing to create a long-term relationship of trust and is likely to undercut customer loyalty as soon as a substitute product becomes available or a competitor enters the market. Second, being the only game in town means there is little pressure to offer proper customer service. Over the long run the take-it-or-leave-it business culture will alienate customers who will be eager to seek alternatives to the products provided by the monopoly. Third, there is little economic pressure to provide top quality goods offered for sale. This attitude often signals a fundamental disrespect for the customer—something that the customer knows and resents. Finally, there is no accumulation of goodwill because the customer relationship is dysfunctional. Much more than merely being dissatisfied with the monopoly (prices, level of service, limited selection of product choices), customers can become resentful. In times of economic downturn, or when a new competitor enters the market, the customer will abandon the monopoly.

As space settlements and outposts are established farther out in the solar system, industry-owned and sponsored "company stores" will likely be the norm. These local monopolies serve a captured customer audience with little or no alternatives available. The company store provides workers with food, clothing, survival gear, entertainment, and other staple necessities, but the high prices charged may exceed the worker' income thus placing the worker in debt to the company resulting in de facto economic bondage. This practice happened with frequency in isolated American frontier logging camps, mining operations, canneries, and other unregulated outposts (Mellon and Wille 2019).

The Impact of Space Innovation on Earth's Economy

Establishing new self-sustaining settlements in space will be much more than pitching an inflatable habitat by the side of a Martian crater. Materials and technologies needed to jumpstart space settlement will likely be repurposed from local resources. This in situ resource utilization (ISRU) may resolve much of the early demand for the basics of living in a new environment. But in time, camping out will give way to more permanent social and economic structures—the beginnings of a settlement commercial economy.

Earth can provide only a limited inventory of equipment and expertise—in due course it will fall to the settlers to determine their own needs and develop solutions that are more in tune with their extraterrestrial priorities. Some of these innovations will be relatively mundane such as housing, communications, waste management, etc. Other more original solutions will probably surpass the current state of practice on Earth. Innovation will replace older, less efficient, operational practices. Some of these new technologies (e.g., for production) may compete directly with Earth-based providers and test the market position of long-established traditional companies. The process of technological diffusion will become a hallmark of the new space economy (Stoneman, 1985). Technological diffusion describes how innovations travel virus-like within and across economies. These new innovations may be manufacturing methods, new products, new processes, new technology solutions to new problems, or even new leadership practices. It is important to remember that just because a novel innovation is successful in one setting there is no guarantee that it will succeed in another. There are several reasons why an innovation may not be quickly adopted in a new commercial setting. High on that list is the cold economic reality of sunk costs due to investments in long-established physical production plants and other capital investments.

Whether diffusion of an innovative technology occurs or how rapidly the diffusion of innovation takes place relies on several key issues. These include the impact the new technology will have on current customers, the prospect of leveraging the innovation to attract a new customer set, the fundamental sustainable quality of the innovation, the perceived value proposition and how to efficiently build a marketing plan to promote the benefits of the innovation, and a solid analytical assessment of the market to see if there is strong economic potential if the new innovation is adopted. When new innovations that are developed out of necessity in space challenge the traditional operational methods on Earth, they are more likely to be adopted for the older company on Earth to stay competitive. This diffusion of innovation is the consequence of strategic decisions about the costs and benefits of adopting the new technology (Hall and Khan 2002). Surrounding these decisions are the issues of return on conversion investments and the uncertain scope of benefits, including customer impact, of the new technology.

Energy generation is an example of the imminent impact of adoption of new innovative technologies created in space. The creation of energy in space for commercial customers who are either in space or on Earth will demonstrate how the rapid diffusion of new inventions will change Earth-based business practices. Among the repercussions of providing space-generated energy to Earth-based customers will be the shift from carbon-based energy to the use of cleaner energy on Earth. Because the energy generated in space can be transmitted over great distances either by microwave or laser, remote impoverished villages and even ships at sea on Earth will benefit along with new settlements and outposts in space (SpaceFund 2020).

Life on the S-Curve

A useful pictorial representation of how the adoption process of the new space economy will progress over time is the S-Curve model. The S-Curve is often used to show the progress of a project, the saturation of a new product in the marketplace, or, in our case, the expected adoption of the budding space-based market. The S-Curve can also be used for a variety of planning purposes to show how a new enterprise grows and develops over time. The S-Curve is employed here to illustrate how industrial space will move from an unorganized collection of bare-boned frontier ventures to a fully functional mature marketplace.

Figure 23.1 shows the process of adoption over time using the S-Curve model. The horizontal bottom axis, the X-axis, represents time. The vertical axis on the left, the Y-axis, may represent various measures from costs, user acceptance, market penetration, revenue, and other dynamic metrics. Here the vertical Y-axis represents expected market maturity. Please also note the two major inflection points represented as arrows on the graphic. These bracket the expected area of growth and development of the venture. In some cases, this may be the point where go/no-go decisions are made if the objectives of the project are not met or if costs of the project are over budget. In some cases, the achievement of an inflection point may signal the necessity for the next stage of additional funding. The slope of the curve in the area between the two inflection points typically represents the growth rate of demand, costs, revenue, etc.

The acquisition of the upper inflection point signals a reduction in growth momentum. At this point the project, product, or new market adoption being measured enters a phase of slower growth on its path to maturity. This phase is usually characterized by increased competition, a commoditization of the product or service, or a saturation level of the general user/customer. Movement along the S-Curve is not a smooth and predictable ride. Like any human enterprise there are fits and starts, conflict and negotiation, budget and funding snafus, and, of course, the innovator's biggest enemy: resistance to change.

Fig. 23.1 This chart plots the dynamic growth of the 5-phase adoption model over time. Superimposed on the 5 phases timeline is an S-Curve illustrating how the space economy will adopt from each phase to the next. Labels of the five phases are adapted from Everett Rogers (1962, 2003).

The early foundational efforts of critical analysis, design, development, and implementation serve as a platform for the later stages on the S-Curve. This means that a firm wishing to enter the market at an early phase would carry a higher development burden but would likely benefit as a provider to all those enterprises who choose to enter later along the S-Curve path. In short: High potential risk = high potential reward. The illustration below is divided into five sections which represent the five phases of development of the new space economy.

Using the S-Curve model is a convenient way to describe how the new space economy will likely develop and mature. Each of the five phases of adoption has its own set of challenges and its own optimal business model (discussed later in greater detail). The cost/benefit analysis that a firm would use to determine the viability of entering the new space economy at a specific phase will probably be a healthy mix of hard data analysis and intuition. Of course, it will probably boil down to answering a few critical questions such as:

- Can our firm take a long-range view and support a long-term return on investment?

- Will our current set of capabilities (talent, technology, culture, secret sauce) mesh with this venture? This is another way of saying: How much do we need to make, buy, or borrow to launch and sustain this new business effort?
- With whom can we partner? With whom can we share the rewards in order to mitigate the risks?
- Do we have a governance model that supports this venture? Will our board be on board for the long term?
- What's the most compatible business model to support this strategic decision?
- What are our pivot options and the costs associated with following a new course of action?

Supply Chains and Service Providers

A benefit of global trade is the ability to dissemble the production process so that labor and material components can be manufactured efficiently by a diffuse network of complimentary providers. The final assembly of a Japanese car may take place in an American factory and consist of parts manufactured in Mexico, Vietnam, and a range of other locations. The supply chain that services this production process is made up of independent companies who constantly compete among themselves for contracts with their auto manufacturing customers (Donnan & Leatherby, 2019).

One of the milestones of a space settlement economy will be establishing supply chains of specialized providers that provide labor (either human or robotic), complex components (telemetry, computerized control systems, etc.), business services (e.g., payroll, accounting, consulting), and all the other elements needed by a space-based enterprise. Of course, each of these suppliers will rely upon their own web of companies to supply them with the components they need to produce their end-product inventories.

A clear bellwether of space market maturity will be the development of a robust web of competitive suppliers servicing a variety of industries in space. The resulting complex network of suppliers and producers will evolve along with the development, diffusion, and adoption of the space settlement economic platform.

References

Amir. (2015, November 11). *Advantages and disadvantages of monopoly*. Economics Guide. Retrieved from http://www.economicsguide.me/?page_id=1044

Donnan, S., & Leatherby, L. (2019, July 23). *Globalization isn't dying, it's just evolving.* Bloomberg.com. Retrieved from https://www.bloomberg.com/graphics/2019-globalization/

Hall, B. H., & Khan, B. (2002). *Adoption of new technology.* Retrieved from https://eml.berkeley.edu/~bhhall/papers/HallKhan03%20diffusion.pdf

Krugman, P. (2010). The theory of interstellar trade. *Economic Inquiry, 48*(4), 1119–1123. https://doi.org/10.1111/j.1465-7295.2009.00225.x.

Mburugu, C. (n.d.). *The benefits of competition—business strategies.* www.gaebler.com. Retrieved June 27, 2020, from https://www.gaebler.com/The-Benefits-of-Competition.htm

Mellon, C., & Wille, D. (2019, August 7). *Space settlements could end up being company towns.* Slate Magazine. Retrieved from https://slate.com/technology/2019/08/space-settlements-company-towns-bezos-musk-the-expanse.html

Rogers, E. M. (1962, 2003). *Diffusion of innovations.* Free Press. (Original work published 1962).

SpaceFund (Crawford, M). (2020). *Energy transmission.* SpaceFund. Retrieved from https://spacefund.com/energy-transmission/

Stoneman, P. L. (1985). *Technological diffusion: The viewpoint of economic theory.* ideas.repec.org. Retrieved from https://ideas.repec.org/p/wrk/warwec/270.html

24

The Rules of the Game

Not everyone is excited about the new space economy. As progress is made and momentum builds for the new space economy, opposition from entrenched business and political sectors will likely come in the form of new restrictive trade policies and regulations or renewed rigorous enforcement of old commerce accords, agreements, and treaties. These policy sanctions may be used as a cudgel to restrict competition instead of as an enabler to promote trade. One measure of their effectiveness is how space companies and spacefaring nations will (or won't) engage in an ongoing review and amendment of original trade pacts. It is hoped a space congress of representatives will work to renew and revitalize these agreements in keeping with the inevitable changes of industrial space.

Space Law and Policy

Since 1967 there have been several well-intentioned attempts to establish international treaties and rules of conduct for nations operating in space. The purpose of these agreements is to ensure a more even-handed approach to space and to reduce the likelihood that space will become the new theater for war. Space policies have been designed to address ownership of celestial turf, the extraction and use of materials found in space, the shared responsibility to not corrupt or destroy the natural space environment, the establishment of shared/common standards for technology compatibility, and the obligation of each nation to assist citizens of another nation who may be in distress.

Individual companies have yet to be invited as a signatory party to any of these documents because it is likely assumed that a company is bound to abide by the laws and commitments of the nation where it is incorporated. This assumption may encourage rogue corporate privateers (pirates) like those of the sixteenth and seventeenth centuries who bore flags of convenience, or no flag at all, as an act of independence. Or they could just incorporate with any willing nation not a party to any space treaty that is keen to merely collect a corporate registration fee. Grand Cayman Island is one such nation of registry convenience that comes to mind because it has neither signed nor ratified the Outer Space Treaty of 1967.

The Space Treaty of 1967, formally the "*Treaty on Principles Governing the Activities of States in the Exploration and Use of Outer Space, including the Moon and Other Celestial Bodies,*" is an international agreement that forms the basis of international space law and was signed by 109 countries. The most often referenced stipulation in the *Space Treaty* is that claiming ownership of all or part of real estate on a planet or other celestial body is not allowed (Wikipedia Contributors 2019). Someone can't go to the Moon, plant a flag, and simply claim it for the motherland. However, it also stipulates that it is just fine for enterprising explorers to locate and extract raw materials for financial gain. This could include palladium on an asteroid or even water located at the Moon's polar regions. This prohibition of ownership but permission to extract materials or scientific knowledge is modeled after the research stations on Antarctica where research has been conducted jointly with other nations for decades at "temporary" sites.

Another attempt at space regulation is *The Commercial Space Launch Competitiveness Act of 2015* (2020) which was enacted to permit American companies to, "… engage in the commercial exploration and exploitation of space resources." It further stipulates, "… the United States does not [by this Act] assert sovereignty, sovereign rights or jurisdiction over, or the ownership of, any celestial body." It may be argued this legislation establishes de facto sovereignty by reserving ownership rights of all materials obtained in space.

A new space directive, formally titled "*Encouraging International Support for the Recovery and Use of Space Resources,*" was signed by President Donald Trump on April 6, 2020. Its purpose is to reaffirm the U.S. commitment to the commercial use of resources in space as outlined in the *1967 Outer Space Treaty*. This directive aimed to repudiate the controversial *1979 Moon Treaty*, which attempted to restrict the exploitation of space resources (Klotz 2020). This agreement was not signed by any major spacefaring nation.

The *Artemis Accord*, drafted May 20, 2020, is an attempt to define the critical values of space exploration in American terms. This is an effort to imprint the future of space with an ethical perspective that is uniquely Western in nature during a time when other nations with non-western ethical perspectives are looking to establish outposts and settlements in space as expressed by the document's introduction stating, "With numerous countries and private sector players conducting missions and operations in cislunar space, it is critical to establish a common set of principles to govern the civil exploration and use of outer space" (Dunbar 2020). Some of the key elements of the Artemis Accord are:

- **Peaceful Purposes.** "International cooperation on Artemis is intended not only to bolster space exploration but to enhance peaceful relationships between nations. Therefore, at the core of the Artemis Accords is the requirement that all activities will be conducted for peaceful purposes, per the tenets of the Outer Space Treaty."
- **Transparency.** "Transparency is a key principle for responsible civil space exploration and NASA has always taken care to publicly describe its policies and plans. Artemis Accords partner nations will be required to uphold this principle by publicly describing their own policies and plans in a transparent manner."
- **Interoperability.** "Interoperability of systems is critical to ensure safe and robust space exploration. Therefore, the Artemis Accords call for partner nations to utilize open international standards, develop new standards when necessary, and strive to support interoperability to the greatest extent practical."
- **Emergency Assistance.** "Providing emergency assistance to those in need is a cornerstone of any responsible civil space program. Therefore, the Artemis Accords reaffirm NASA's and partner nations' commitments to the Agreement on the Rescue of Astronauts, the Return of Astronauts and the Return of Objects Launched into Outer Space. Additionally, under the Accords, NASA and partner nations commit to taking all reasonable steps possible to render assistance to astronauts in distress."
- **Registration.** "Registration of space objects is at the very core of creating a safe and sustainable environment in space to conduct public and private activities. Without proper registration, coordination to avoid harmful interference cannot take place. The Artemis Accords reinforces the critical nature of registration and urges any partner which isn't already a member of the Registration Convention to join as soon as possible."
- **Data Release.** "NASA has always been committed to the timely, full, and open sharing of scientific data. Artemis Accords partners will agree to follow NASA's example, releasing their scientific data publicly to ensure that

the entire world can benefit from the Artemis journey of exploration and discovery."
- **Heritage**. "Protecting historic sites and artifacts will be just as important in space as it is here on Earth. Therefore, under Artemis Accords agreements, NASA and partner nations will commit to the protection of sites and artifacts with historic value."
- **Resources**. "The ability to extract and utilize resources on the Moon, Mars, and asteroids will be critical to support safe and sustainable space exploration and development. The Artemis Accords reinforce that space resource extraction and utilization can and will be conducted under the auspices of the Outer Space Treaty, with specific emphasis on Articles II, VI, and XI."
- **Deconfliction of Activities**. "Avoiding harmful interference is an important principle of the Outer Space Treaty which is implemented by the Artemis Accords. Specifically, via the Artemis Accords, NASA and partner nations will provide public information regarding the location and general nature of operations which will inform the scale and scope of 'Safety Zones'. Notification and coordination between partner nations to respect such safety zones will prevent harmful interference, implementing Article IX of the Outer Space Treaty and reinforcing the principle of due regard."
- **Orbital Debris and Spacecraft Disposal**. "Preserving a safe and sustainable environment in space is critical for both public and private activities. Therefore, under the Artemis Accords, NASA and partner nations will agree to act in a manner that is consistent with the principles reflected in the Space Debris Mitigation Guidelines of the United Nations Committee on the Peaceful Uses of Outer Space. Moreover, NASA and partner nations will agree to plan for the mitigation of orbital debris, including the safe, timely, and efficient passivation and disposal of spacecraft at the end of their missions."

If there's one lesson learned from watching Gary Cooper, Randolph Scott, or John Wayne in their iconic roles as Western movie heroes, it's that the rule of law means nothing on the frontier unless there's a strong sheriff willing to keep the bad guys in check. My [admittedly cynical] view of the best intended space law is that opportunists will very likely take advantage of others when there's no threat of sanctions. Keeping the peace or enforcing laws on the vast uncivilized range of the space frontier will be next to impossible for a long time. So, you better come well prepared.

The Foreboding Land of Regulations

If the idea of sending people into lethal space wasn't bad enough, then, just for fun, add a heavy dose of bureaucracy and regulation to make the whole enterprise even harder. At the top of the list of regulations is ITAR. The ***International Traffic in Arms Regulations*** (ITAR) is a U.S. set of rules and policies originally designed during the cold war when American allies were threatened by Russia's Warsaw Pack alliance (Lord 2015). The idea of ITAR was to keep proprietary and classified technologies used for defense-related technologies out of the hands of the enemy. Because some of these technologies have dual military and civilian use navigating ITAR has been a challenge. ITAR is more than a way to protect design IP; it ensures that every part and component of every defense-related piece of equipment will be made by a handful of allied nations (like Japan, Britain, and Israel). Further, only an approved list of companies operating in these approved countries who have appropriate clearance classifications can be recognized ITAR suppliers. This is a logical approach from the perspective of national security. But it turns out there is much about the technologies of space and defense that overlap (think: missile technology, satellite sensors, data analytics, sophisticated onboard computers, etc.).

The logic of ITAR dictates that if we were to go to war with an adversary and that enemy nation also supplied key components for our military equipment, then we would no longer control the safety of the supply chain nor the quality of parts in that supply chain. This makes good sense. A current case in point is the international conflict concerning using Huawei Technologies equipment for 5G and other leading-edge technologies. Huawei is a wholly owned entity of the Chinese government thus causing great concern in the west about the not-so-subtle incursion of sensitive technologies into the military supply chain from an economic (and political) adversary like China.

The effects of ITAR restrictions on the U.S. space industry are substantial. As space has shifted from the government to the private sector, ITAR regulations have remained in place. Everything from nuts and bolts to state-of-the-art electronic components must be approved for use by the American defense industry and, by default, much of the U.S. space industry as well. ITAR also applies to personnel: only U.S. citizens or Green Card Aliens are authorized to work on classified projects. The good news is that ITAR serves the interests of national security and ensures that all the components that go into defense systems meet high production standards. The bad news is that when the purchasing process and competition are limited due to the restrictive nature of

ITAR, then prices tend to increase which results in high space hardware costs. High priced U.S. space hardware means foreign space firms can compete more readily in the global commercial space marketplace. The U.S. Department of State has dismissed concerns voiced by the private sector about the negative impact of ITAR on competition by citing the importance of national security over economic interests.

The idea behind ITAR is sound; it makes good sense to acknowledge that technologies used for national security need to be protected and rules for violating the policies governing the ITAR firewall certainly need to be enforced. But until commercial and defense-related imperatives can be fully reconciled, the adverse impact of ITAR on U.S. space companies will only serve to encourage these companies to move offshore or to not adapt technologies developed for space for defense applications. Either way it is a lose/lose outcome.

Your New Settlement Home Is the First Star to the Right, Then Straight on "Till Morning"

As settlements steadily spread throughout the solar system, new markets and opportunities will grow along with them and continue to fuel a vigorous space economy via trade and industrial investment. The dynamism of this growth will come from a continuous array of new innovations. These will produce products and services to satisfy market demand from individual consumers and corporate customers. The resulting growth will supply investors and entrepreneurs with an endless commercial arena for new economic opportunities.

Frederick Jackson Turner, writing about the closing the American frontier in 1893, saw the end of western expansion as a harbinger of social and economic disaster. Writing before the electrification of homes and cities, the age of air travel and automobiles, the upheaval of world wars, the new deity of consumerism, and the new wealth from escapism and social connection of radio, talking pictures, television, and the internet, Turner saw the end of the frontier as a sign of approaching equilibrium. He feared the dawning of new age of supply and demand would reach a state of stasis, the end of the trail for western economic growth. He guessed that the pending market maturity was proof that the apex had been reached and passed; that the rest of the ride for civilization was all downhill.

He was correct about the end of the American frontier and the social impact of no more free range for the outcasts and iconoclasts of society. But he overstated the consequences—the end of the physical frontier wasn't the dramatic end of civilization's forward march. Looking back to the late nineteenth century, it is plausible to see that since we had come so far so quickly that there was little left to improve. But it turned out that innovation kept making room for new industries that served more customers and employed more labor. Boom and bust may have caused the economic roller coaster rides of 1907, 1929, and 2008 (Hayes 2019) but following each recession came increasing market efficiencies, corporate consolidations, protective financial regulations, and technical innovations that provided individual consumers and business customers with more products and services and created new markets that Turner would not have recognized.

If industrialization of the solar system continues to expand because of a seemingly endless formation of frontier outposts, space settlements, and far-flung communities in space, then Turner's dire predictions will not come true. The classic notion of mature market equilibrium, where supply perfectly balances demand—where economic growth comes to a halt, will simply not happen. In this perpetual impetuous adolescent state, just short of market maturity, the gap between financial aspiration and satiation will continue to energize the space economy. For those seeking their ideal new home in space it may be found if you dare venture to go to, "… the first star to the right and then straight on 'till morning" (Barrie 1911, 2019).

References

Barrie, J. M., & Bedford, F. D. (1911, 2019). *Peter pan*. Suzeteo Enterprises. (Original work published 1906).
Commercial Space Launch Competitiveness Act of 2015. (2020, May 24). *Wikipedia*. Retrieved from https://en.wikipedia.org/wiki/Commercial_Space_Launch_Competitiveness_Act_of_2015
Dunbar, B. (2020). *NASA: Artemis accords*. NASA. Retrieved from https://www.nasa.gov/specials/artemis-accords/index.html
Hayes, A. (2019). *Boom and bust cycle*. Investopedia. Retrieved from https://www.investopedia.com/terms/b/boom-and-bust-cycle.asp
Klotz, I. (2020, April 7). *New space policy directive supports commercial space mining rights*. Aviation Week. Retrieved from https://aviationweek.com/shows-events/space-symposium/new-space-policy-directive-supports-commercial-space-mining-rights?utm_rid=CPEN1000002530288&utm_campaign=23673&utm_medium=email&elq2=4cb42cf1ee484b42af9de10349e9767b
Lord, N. (2015, March 30). *What is ITAR compliance? 2019 ITAR regulations, fines certifications & more*. Digital Guardian. Retrieved from https://digitalguardian.com/blog/what-itar-compliance
Wikipedia Contributors. (2019, February 6). *Outer space treaty*. Wikipedia; Wikimedia Foundation. Retrieved from https://en.wikipedia.org/wiki/Outer_Space_Treaty

25

Trouble in Paradise

At some point in the very near future the dream of human space settlement will either thrive and become the next step for human civilization or die and become another lost cause, a shriveled dead branch on the tree of historical possibilities. Dreams of space settlement and economic expansion have waxed and waned before. After Armstrong and Aldrin landed on the moon in July of 1969 there were high hopes that this amazing accomplishment was only the first step in a grander plan of further space exploration and settlement: an era of Moon bases, permanent human occupation, the seeds of a cosmos-based economy. The dream fizzled when the funding ran out along with the political will to fight for the dream. The Moon shot was only a political statement to demonstrate that American technology and commitment was superior to that of Russia. After we celebrated our victory in the space race we retreated to other, safer, national priorities that redirected space expenditures to Earthly concerns and stranded the dream of human space settlement back out on a limb.

In spite of this, the impact of Apollo 11's success is still strong enough in our collective national memory to know the taste of accomplishment, the joy of discovery, to feel the satisfaction that all the effort and coordination worked as planned: that a human from this planet could travel to another celestial body, hit a golf ball, and come back alive. It is this foundational memory that inspires the many space entrepreneurs seeking to plant their personal flag on some extraterrestrial rock in the cosmos and move us all one step closer to an era of space settlement.

Today China and other foreign nationals have jumped into the gap created by the American retreat and are mustering their resources to establish their space primacy. In response, American private enterprise has countered with leadership and bold plans for space settlement. Space-based capital ventures like mining, manufacturing, and other profit-seeking enterprises are attracting venture capital to fund enterprises in space. The new space race may not be between countries but among private companies who seek robust ROI instead of mere political points.

There are those who see efforts to build a space economy as the wrong path. They argue that funds spent in pursuit of this agenda are better spent on Earth-based projects like infrastructure improvements, social services, or military development. Going to space—creating a new space economy—isn't a matter of either/or. It isn't a zero-sum game where there is a winner and a loser. It is a matter of *and*, a win/win game where both agenda can thrive. Consider the consequence of abandoning the agenda of settling space. Why would we choose to forego this opportunity? Who would benefit from this retreat? Would the momentum of creating a commercial space economy ever be rekindled if the dream were extinguished?

The inflection points (recall the S-Curve) when a space settlement economy is most likely to be rejected or thwarted will likely occur (1) during the first few phases of commercial adoption of the frontier space-based economy, and (2) at the decision point to redirect financial support to more lucrative commercial markets on Earth rather than develop a new autonomous space economy. Those in opposition to funding and nurturing a new space economy will likely be established companies and nations who view the developing space economy as an imminent threat to their power, influence, or profitability. For naysayers the onset of a new commercial space economy would potentially drain revenues, redirect capital investments, and create competitive opportunities for new entrants that would challenge their existing market position. They would have three strategic options: (1) fight the efforts to create a new space economy, (2) accept that space will become a significant competitive addition to Earth's future, or (3) join the effort and become part of the new space economy.

The argument to forego space investment in favor of social, political, or military investments will be made by companies and countries with an established commercial platform who may see the potential of the new space market as a dangerous undertaking. It may be more expedient for well-established firms to opt for business as usual, to put their energy and resources toward efforts to maintain and buttress their market position, and to ardently resist the threatening new space economy. It is also likely that enough firms will see the new space business setting as an epochal opportunity, like the new digital business opportunities that arose from the advent of the internet.

Along the way there will likely be a series of ups and downs as there is for any market arena. There will be periods of consolidation, changes in brand dominance via corporate merger and acquisition activities (M&A), and a long list of companies who will disappear due to underfunding, inadequate management, a flawed business plan, an inability to pivot as the economic environment shifts, or a host of reasons that have stalked entrepreneurial start-ups forever.

Of primary concern is how the leadership team of an entrepreneurial space organization will deal with the uncertainties and the ambiguous nature of adopting to the new space environment. Where the business context on Earth is relatively mature and predictable, the space-based venture will have to contend with the usual day-to-day issues of making payroll, repairing robots, and keeping the lights on. An early venture in space will have to learn to respond to the constant changes in a volatile economic setting such as the transition between adoption phases or the constant operational investments necessary to keep the business vital. The best advice to give a manager of an early phase 1 business is to anticipate that change will happen and to be ready to reassess assumptions about customer demand, product, price, marketing promotion, distribution, and the critical choke points in the supply chain.

When scanning the playing field for potential competitors or threats it may be just as important to identify future market trends as to consider complacency as the worst enemy a manager may have; it will undermine and threaten the long-term sustainable competitive advantage of the operation. When things start to get too comfortable, it is time to consider making changes. The new motto in the volatile setting of commercial space may be, "If it ain't broke, then fix it anyway!"

Part V

The New Space Economy: Setting the Stage for the Industrialization of the Solar System

This section acknowledges the potential opposition to the development and adoption of the new space-based economy, discusses how creativity and innovation will enable space industrialization, presents best case business models for those enterprises planning to establish operations in space, presents the five-phase model of adoption of the space economy, and finally shows how business sectors, business models, and phases of adoption come together in a unified model.

26

Industrial Space

Creativity and Innovation

Adapting to new markets is not a simple undertaking. Harvard Business School Professor Clayton M. Christensen (1952–2020) famously noted in *The Innovator's Dilemma* (1997) that well-meaning companies could stumble and fail to innovate in a rapidly changing business environment by doing two wrong things even though these things are widely taught as right things in leading business schools. They are (1) listening to their best customers and taking their advice for product feedback and improvement suggestions and (2) focusing investments on innovative initiatives that promise to deliver the highest immediate returns. Christensen argued that both long-standing B-School tropes could spell disaster by perpetuating the *current* product and customer base instead of seeking ways to serve the *future* market. He insisted that instead of getting comfortable feedback from current customers who reinforce the old way of doing business it is best to seek out new input from the leading fringe of the market.

Note that the terms creativity and innovation are not interchangeable. Creativity refers to the ability to generate new ideas or solve problems in new ways. Innovation is the ability to put new ideas to use. Innovation is implementing the creative idea in the marketplace. Or, as the Harvard economist and former editor of the *Harvard Business Review*, Theodore Levitt said, "Creativity is thinking up new things. Innovation is doing new things." (2002).

Professor Mark Allen of Pepperdine University has consulted and lectured on the important strategic distinction between creativity and innovation. He presented his research on this topic to my business students (April 2020) concluding that organizations with open climates have no shortage of creative ideas while organizations with closed and restrictive climates suffer from an inability to execute, to take creativity to the marketplace, and thus they fail to hold their own in a competitive market.

Firms with well-established customers, dedicated supply chains, and fixed brand identities become wedded to doing business comfortably and may adopt a corporate culture of inflexibility. It is the tried and true processes, after all, that got them to where they are and defined their competitive success. To make a radical modification takes a conscious effort to transpose creativity into implementable innovation. That means organizational change, an activity which notoriously evokes resistance from a full range of stakeholders. Kurt Lewin's *Field Theory* (1951) represented as "b = f(p,e)," defined how change could be managed. Lewin explained how **b**ehaviors "b" were the **f**unction "f" of the individual manager's **p**ersonality "p" and the organization's **e**nvironmental context "e" (Burnes 2004; Raza 2019).

One of the classic elements of Dr. Christenson's book is the radical notion that to make a transformative change, to do something that is completely outside the pale of the generally accepted corporate practice, a leader must make a leap of faith and give up the safety of the current customer base. That is not easy. Customers are the hard-won life blood of success not just because they provide the revenue that generates corporate financial health and because a firm's current customers are also the key relationships that connect the enterprise with the greater marketplace and even to competitors. The common process of making incremental product improvements has always been to talk with major customers and respond to their concerns.

Christenson challenged this. He said that to do something that is truly innovative, truly transformative, the innovative firm should NOT talk to their customers. That is because of something called a "selection problem." It occurs when the data used to make a decision is collected by talking only with like-minded respondents. But they are not an objective sample of the broader market. Current customers are biased to the status quo—they already purchase the company's product and thus don't seek changes. Asking them to suggest innovations won't reveal *objective* transformative opportunities for innovation that a less biased sample would provide. There's nothing ill-intentioned, it just doesn't get the job done because the current customer base is not an objective random sample. If a company insists on talking only with current customers to figure out how to be innovative, they will never learn anything other than how to make small, incremental, gradual changes to the current product line.

Real innovation is much more than fine-tuning products around the edges. True innovation is taking a giant step, a leap of faith, which embraces high risk in hopes of gaining high reward. True innovation is ripping up the old assumptions that were used to successfully define success before and starting with a blank sheet of paper to make up a whole new game plan. True innovation is a very scary thing, which is why most firms step back from the edge of the cliff at the last minute. For this reason, Dr. Christensen's words about risk and innovation are prescient when considering economic adoption on the new space frontier.

The Frontier and Colonial Commerce

During the European expansion into the new world several noteworthy attempts were made to attract investors to spread economic risk and to develop substantial return on investment (ROI)). A few notable examples include The Virginia Company, The Hudson Bay Company, the Dutch East India Company, and The South Seas Company famous for lasting only 10 years (1711 to 1721) due to mismanagement and inflated claims.

The Virginia Company (founded 1606) was created as an investment enterprise of London Merchants using the structure of a joint stock company to combine their investment in order to purchase stock in the enterprise. When Britain's King James I decided to create new settlements in the colony of Virginia, he established the organizational structure of a joint stock company that encouraged individual investors. The Jamestown experiment (1607) was ultimately unsuccessful and ended in disaster, but because the risk was spread among a group of investors, King James didn't suffer the burden of loss all by himself. The Jamestown experiment is significant because it demonstrated the benefits of mitigating risk for future investment schemes.

The Dutch East India Company (established 1602) is the world's first truly transnational corporation. It was the first company to issue bonds to raise capital and shares of stock to expand ownership to the general public. Selling corporate debt and equity increased access to capital just as the aristocracy and upper middle class were accumulating wealth due to the financial impact of the industrial revolution. Another example of early investment companies that sought profit in the new world is The Hudson's Bay Company. It was originally incorporated by an English royal charter in 1670 under the name of "The Governor and Company of Adventurers of England trading into Hudson's Bay." Over the ensuing centuries it became more than just a company and rose to become the de facto government in parts of Canadian North

America before European states and later the United States laid claim to some of those territories. In fact, at one time it was the world's largest landowner. After nearly three-and-a-half centuries it was dissolved in 2012 and has morphed into a retail brand in Canada.

The danger of companies that act like states is that there is a tendency to blur the boundary between being a public country and private company because their autonomy affords them potential unchecked power over resources, territories, legal systems, and the people who live and work under their control. Phillips and Sharman, writing on this topic in their book, *Outsourcing Empire: How Company-States Made the Modern World* (2020), noted that the East India Company had a private army of more than 250,000 soldiers that were used to suppress dissent and impose company-mandated policies and laws. At one time Hudson's Bay Company had complete control of nearly one-tenth of the planet's land mass and enforced political and economic influence accordingly.

These powerful hybrid company-states are the first true multi-national organizations. During the seventeenth and eighteenth centuries when they thrived unchecked, their power and worth exceeded that of today's leading global firms. Their virtual free reign in their pursuit of profit and power superseded sovereign control in a time when international law was really a matter of international custom among royal monarchs. As a result, there was no accountability for their actions regarding the treatment of people or the exploitation of local resources and little accountability to their investors and stockholders.

Many of the early investment companies that were initially organized to profit from the economic frontier of the New World survived into the twentieth and even the twenty-first centuries which is some indication of their success. Similar investment models are likely to be organized to raise capital and profit from the emerging space economy.

The often-used entrepreneurial buzzword today is flexibility. Business plans are no longer expected to cover every eventuality whether anticipated or not. The ability to "pivot" to respond to the influence of new technologies, to unproven business assumptions, to create a new roadmap from the ashes of the old business plan are all now accepted as business as usual when launching a new venture. Another word for enterprise flexibility is opportunism.

Consider the California Gold Rush of 1849 and the adventurers who came seeking instant easy riches but became entrepreneurs instead. They, and others who came west, saw something more than gold coming from the placer mines in the mountain streams above Sacramento. They recognized something special in the new towns springing up throughout the new American western frontier: commercial opportunity by selling goods to the new arriving immigrants. They saw demand for their products, took advantage of the moment, and built their businesses along the way as the frontier morphed and adopted to more mature and robust markets.

Levi's became a worldwide garment brand. Studebaker (1852–1967) started as a wheelbarrow maker catering to gold miners, pivoted to building rugged wagons, and years later launched a car company. Rural peddlers like Goldwater, Sears, Woolworth, and Montgomery Ward, took merchandising directly to their customers—the original Amazon-like convenience value proposition—with catalog sales and postal home delivery.

Managing Space

Current notions of management (e.g., planning, accountability, command, and control) are based on the nineteenth century military and early industrial models of manufacturing and production. These antique modes have given way to modern notions of employee motivation, engagement, alignment, and empowerment—something unheard of until recently. What will managers need to know to be effective in a post-industrial space economy where AI-driven robotic workers will function at all levels of the future firm? Compound this with the notion of a workforce made of both human and non-human contributors and the role of the manager becomes more complex. In a future work world without people to gum up the works the traditional manager's role may be reduced to planning and controlling the supply chain and maintaining the mechanical health of the robotic workers.

For millennia trade and commerce have been characterized as transactions of profitable exchange between a producer and a consumer. As markets became more complex middlemen, or agents, emerged to distribute goods from producers to other middlemen and then finally to customers. It was up to the people at the end of this distribution chain to build and maintain customer relationship and provide feedback and input back up the chain to the producer about changes in demand, competitor threats, or issues of quality. Whether selling directly to an end consumer or through an intermediary agent the role of the personal relationship with the customer, the end user, was always the bedrock of sustainable trade. The emergence of mass production and simplified distribution removed the middlemen and closed the gap between producer and customer causing the personal relationship with the consumer to change. Now producers can directly connect with thousands of customers. In this new setting the once sacred customer relationship has devolved to simplistic transactional criteria such as cost, quality and time to fulfillment.

This is evidenced by the steady decline of the traditional retail sector (e.g., Macys, JC Penney, Nordstrom, Sears, and others) that has been eclipsed by the direct and more convenient transactions between seller and consumer via the internet (e.g., Amazon, Walmart). Eliminating the layers of transactions between producer and consumer, a process called *disintermediation*, cuts costs and thus selling price which makes it very attractive to the end user. Technology-enabled trends that encourage disintermediation have facilitated a more direct connection between the consumer and the producer, thus creating a more efficient marketplace of goods and services. The removal of distributors and other intermediators has also caused a rejiggering of the logistical structure of how retail has traditionally functioned. And it has eliminated much of the human side of the commercial equation.

There is no evidence to believe that the trend to bring producer and consumer closer by removing layers of distributors and wholesalers will change appreciably as production, commerce, and consumption expands beyond Earth to new outpost markets in space. The same fundamental transactional models will continue to prevail into the near future. Whether or not the role of the personal relationship in facilitating these transactions will persist is unknown. However, given the impact of "big data" on reaching, coddling, and cajoling micro markets of small consumer cohorts, it is expected that this new form of relationship marketing will spawn customized manufacturing and production for both tangible and intangible products.

Adopting to the Next Phase of Development

The new space economy will advance in a successive series overlapping phases of adoption. The term adoption means acceptance of the new economic setting; acceptance of companies doing business, acceptance of customers, clients, and consumers about the economic realities of a phase, and acceptance of all the interlocking elements of the value chain. Each phase of adoption will offer a set of enabling attributes for an investor that may, or may not, best fit the investor's business model. It's a matter of timing. Table 26.1 offers a summary view of the five phases of adoption. This chart is meant to set the stage for a more comprehensive discussion later.

Each phase of adoption represents economic activity enabled by and built upon the economic developments that came before. Like building blocks. Late Mainstream adoption (phase 4) is shaped by the successful adoption of

Table 26.1 In this table the phases of adoption are shown sequentially. However, when we consider the immense domain of space it is highly likely that different phases may be adopted faster than others or may even progress in a non-linear way. Anticipated date ranges are approximate and intentionally overlap.

Key Characteristics of Adoption Model*

	Name	Cumulative Adoption %	Independence From Earth	Investment Risk	Anticipated Adoption Years
Phase 1	Innovators/ Frontier	2.5%	Low Independence	Highest Risk	2030 - 2060
Phase 2	Early Adopters	16%			2050 - 2075
Phase 3	Early Mainstream	50%	Moderate Independence	Moderate Risk	2070 - 2095
Phase 4	Late Mainstream	84%			2090 - 2115
Phase 5	Lagging Adopters	100%	High Independence	Lowest Risk	2105 →

* Adapted from Everett Rogers *Diffusion of Innovation* (1962)

the Early Mainstream (phase 3), the Early Adopter phase (phase 2) and by the Innovation phase (phase 1). Collectively these earlier phases will set the stage for a more stable phase 5—Lagging Adapters. The transition of moving from one phase to another is unlikely to be simple or linear. Once the development of space has begun there will be a variety of efforts to establish outposts and settlements, conduct mining and refining operations, and manufacture goods from materials harvested in space. Many of these commercial activities will follow different formulae for success (i.e., business models). Some efforts, because of the fundamental nature of their business plan, will remain at an intermediary phase on purpose in order to exploit and maximize the unique benefits of that phase.

Developing innovative products, enlisting eager investors, and planning market entry with the suitable phase of adoption are only part of the formula. The next step that converts planning into action is to define the right business model.

References

Allen, M. (2020, April). *Creativity and innovation* [Guest speaker presentation]. Organizational Behavior Class, Whittier College.

Burnes, B. (2004). Kurt Lewin and the planned approach to change: A re-appraisal. *Journal of Management Studies, 41*(6), 977–1002. https://doi.org/10.1111/j.1467-6486.2004.00463.x.

Christensen, C. M. (1997, 2016). *The innovator's dilemma: when new technologies cause great firms to fail.* Harvard Business Review Press. (Original work published 1997).

Levitt, T. (2002, August). *Creativity is not enough.* Boston, Harvard Business Review. Retrieved from https://hbr.org/2002/08/creativity-is-not-enough

Lewin, K. (1951). *Field theory in social science.* Harper & Brothers. (Original work published 1951).

Phillips, A., & Sharman, J. C. (2020). *Outsourcing empire: How company-states made the modern world.* Princeton: Princeton University Press.

Raza, M. (2019). *Lewin's 3 stage model of change explained.* BMC Blogs. Retrieved from https://www.bmc.com/blogs/lewin-three-stage-model-change/

27

Business Models

A business model is the fundamental *structure* of the enterprise. The business plan is how the company uniquely decides to execute and provide value to its customers. Once the business model is defined and understood by the core constituencies (vendors, customers, competitors, employees, government regulators, investors, etc.) then the business plan can be defined, implemented, and measured.

There is a distinction between the business model and the business plan. The effectiveness of the business plan, how the enterprise will generate revenue, is dependent upon the sound foundation structure of the business model (Ovans 2015). In fact, one way to measure the success or failure of a company is how the business plan conforms to the underlying business model. The importance of a strong business model is that it defines how an organization creates, delivers, and captures a long-term sustainable and competitive advantage in the market. A viable business model is based upon the core assumptions of the environment. The assumptions about the American Western frontier produced business models that were unique to that time and place. Assumptions about the new space environment will help to define the optimal business model choices.

Note that these assumptions are based on the idea that independent pioneer families were responsible for most of western settlement while it is anticipated that corporate actions will drive space commerce and settlement. The western frontier was dependent upon cheap and plentiful human labor, but space will depend upon AI and robotic technologies to do the heavy lifting. In both cases transportation represents a high cost.

Baked into the business plan is a clear understanding of what unique, or innovative, product or service is being provided, who the core customer is, how the product or service will satisfy the customer's needs, what resources (e.g., technologies, human talent) are required to service the plan, and what costs can be controlled to sell the product or service at a profit. Different business models will shape how a business plan is crafted, executed, and measured.

Not all businesses operate the same way nor expect to make money the same way. According to Harvard's Dr. Joan Magretta (2014), a business model consists of two parts.

> Part one includes all the activities associated with making something: designing it, purchasing raw materials, manufacturing, and so on. Part two includes all the activities associated with selling something: finding and reaching customers, transacting a sale, distributing the product, or delivering the service. A new business model may turn on designing a new product for an unmet need or on a process innovation. That is, it may be new in either end.

A business model is the way a company is structured to make and sell goods or services to satisfy customer needs, it's the secret sauce that differentiates one company from its competitor. It is how a company becomes successful and sustains its commercial activity (or not). But business models don't last forever.

The traditional retail business model, where customers come to a store to chat with a favorite salesperson, examine products, and make purchases is no longer as viable as it used to be and is falling out of favor. Retail shops used to dominate how goods were sold until Amazon came along. Shopping was accepted as recreation, a chance to spend time outside of the home experiencing new encounters with people and products, trusting relationships were often developed with salespeople who would make recommendations. Shoppers were willing to accept the inconvenience of large crowds and inconvenient parking lots to engage in the shopping experience. The recent Covid19 pandemic (2020) demonstrated how a traditional business model that relied upon intimate human interaction can be quickly and unexpectedly disrupted and nullified. When shoppers realized they could get grocery and restaurant deliveries to their home they eagerly embraced the new business model based upon convenient home delivery. This switch upended the restaurant and grocery business model that was dependent upon last-minute impulse shopping either at the store or when upsold additional menu items by the restaurant server (would you like fries with that?). Other examples of business model disruptions include:

- Healthcare. The old "cost-plus" business model, where physicians and hospitals charged whatever they wanted for medications and procedures, changed drastically in the new era of managed care and will see a further overhaul in the years ahead.
- Higher education. The move to internet-based distance learning has made it easy for students to virtually attend class at their convenience. This has also removed the added value of an in-person college experience which has caused colleges to cut tuition costs and eliminate jobs.
- Manufacturing production. Just-in-Time (JIT) production traded stockpiling materials for daily deliveries from suppliers. This reduced overhead costs (labor, rent, insurance) and freed up capital previously tied to large inventories. The 2020 covid19 pandemic prompted consumer hoarding, but retail grocery outlets who implemented JIT couldn't keep up with demand and their shelves emptied because they no longer stocked inventory beyond a few days' supply.
- Stock brokerage. The archaic business model of paying high fees to a stockbroker for buying and selling a stock has been replaced with free trading platforms provided notably by Fidelity, Schwab, and others. Independent traders can get all the free research they need online, thus circumventing the need to deal with commission stockbroker sales agents.
- Travel. Taking a trip used to require a travel agent who extracted hefty commissions, but the travel business model was replaced with online services. Why use a travel agent to book your hotel, car rental and air when you can manage it yourself? In fact, why book car rental when you can just grab a Lyft or Uber at the last minute? (Note: Business models can be disruptive. Hertz car rental company filed for bankruptcy in 2020 due to the impact of ride share companies like Uber and Lyft.)
- Bookstores. Even smaller communities were able to sustain independent bookstores. It was common for a trusted store clerk to make a book recommendation to a regular patron. The convenience of Amazon erased that business model based upon personal relationship.

By way of illustration, here are a few business models likely to fit well in the different phases of economic adoption in space. There are many more business models, some old and some not yet fully formed. This short inventory will present a fair idea of how a business using a given model thinks about acquiring, serving, and retaining its customers. This is not meant to be a comprehensive primer but, rather, an overview of some selected business model examples.

To refresh your memory about the five-phase adoption model you may refer to Table 26.1 at the end of the previous chapter. The five phases are the Innovation phase (phase 1), the Early Adopter phase (phase 2), the Early Mainstream phase (phase 3), the Late Mainstream phase (phase 4), and the Lagging Adapters phase (Phase 5).

Examples of Common Business Models

Table 27.1 lists ten common business models that are likely to support business activities in the new space economy. The figure indicates which phase of economic adoption is a likely best fit for each of the business models. The rest of the matrix matches each business model with four provider categories (sovereign state, corporate, private investor—including venture capital and private equity, and autonomous settlements). The new space economy will attract

Table 27.1 This table shows the relationship of a few selected common business models along with (a) likely optimal adoption phase(s), and with (b) a set of commercial entities likely to employ the business model for each phase(s) of adoption in the space-based economy

	Selected sample business models	Likely phase(s) of adoption	Sovereign nation	Corporate entity	Investor/ VC/PE	Settlement (autonomous)
1	SOE	1–3	X			
2	PPP	1–3	X	X		
3	Long tail	2–4		X	X	
4	Open source (strategic relations)	2–5	X	X	X	
5	Traditional exchange	2–5	X	X	X	
6	Vertical integration (captive customers)	2–5	X	X		X
7	Lease/rent only	2–5	X	X	X	
8	Lease/rent to own	3–4		X	X	
9	Franchising	4–5			X	X
10	Platform	4–5		X		X

Notes: (1) The likely phases of adoptions are shown as a range, (2) the four entities represented here are meant to be providers/producers, not consumers or customers, and (3) the term "settlement" represents all manner of independent space communities and outposts

entrepreneurs and investors eager to profit from the emerging commercial setting. Some of the business models that will gain traction are focused on an immediate short-term return and others take a longer-term expectation of return on investment.

Shifting from Pipes to Platforms

Technology analyst, Sangeet Choudary (2013), writing about why business models fail, contrasts *pipes* (linear transactional business models) with *platforms* (networked transformational business models). In the case of pipes, firms create goods and services, push them out and sell them to customers. Value is produced at one end of the pipe and is consumed at the other end of the pipe. There is a linear flow, much like water flowing through a pipe.

Unlike pipes, platforms do not just create stuff and then push it out to the marketplace. The prevailing platform model relies on two different sets of customers. One set of customers pays for advertising or for subscription and the other typically gets content for free while enduring all those ads purchased by the other set of customers. There is a synergistic flywheel between the platform provider and the two sets of customers. The advertising rates are based upon viewership (the volume of visits to the platform) and viewers are attracted to the platform because of the content it provides (e.g., education, entertainment). A platform creates customer value by acting as market exchange where multiple consumers and producers maneuver for shared value. While the platform business model has become one of the dominant models of the current twenty-first century it works best in a mature market where there is a robust community of providers and consumers. The frontier space economy will not likely have such a healthy playing field early on. The general transformation from a pipe model to a platform model is currently disrupting a broad swath of industrial sectors and may well migrate to space along with many new entrepreneurial enterprises seeking to establish a foothold.

There are a variety of business models and strategies that a firm may use as a foundation for operational success. A quick scan of the literature yields dozens of constructs ranging from the notorious Ponzi Scheme to a simple brick and mortar retail operation. In recent decades, there have been new developments that have influenced how companies pivot away from old assumptions in order to embrace new ways to connect with customers. These advances include the introduction of the internet, the reduction of intermediaries in the distribution pipeline (aka: disintermediation), and the pendulum swings between centralized and decentralized approaches to production and distribution, to name a few.

A half-dozen business models are presented here in greater detail. These include the public–private partnership (PPP), the State-owned Enterprise (SOE), vertical integration, the open source business model, the long-tail business model, and the traditional business model. Each approach supports viable business plans well-suited for one or more of the five adoption phases discussed a bit later. These few models are presented only as suggestions to prompt thoughtful consideration by a firm or investor planning to enter the space economy or for investors to help evaluate potential success.

PPP Model: Think—Railroad and Government Financing Deals

We start with and spend more time with the public–private partnership (PPP) business model because it is the funding model most often cited when the question arises about how the future of the cosmos economy will be capitalized. The PPP model was the default model used by NASA for decades.

Public–private partnerships (PPP)) have been in existence for centuries. The proposition that Columbus pitched to Queen Isabella of Spain was essentially a partnership of "public" funding from the Crown and private expertise and execution from Columbus and his colleagues. Simply put, a PPP is a contract between a public sector authority (government entity) and a private party (or investment group). Typically, the cost of the service is borne exclusively by the users of the service and not by the taxpayer which reduces the political risk on the downside in case things don't work out as expected.

A public–private partnership enables the public sector to harness the expertise and efficiencies of the private sector without having to make a permanent commitment of creating this special capability as part of the government bureaucracy and budget. It may be structured so the public sector entity that is seeking a capital investment does not incur any borrowing costs. Instead, PPP borrowing is incurred by the private sector partners and guaranteed by the public agency partner. There is no guarantee of additional funds if future financing is needed.

One of the benefits of the PPP model is that the partnership often enables dual use of capital assets created by the partnership. Civilian/military shared use of Cape Canaveral, the Interstate Highway system, Military airfields that host commercial facilities, and shipping ports and terminals that accommodate both civilian and military vessels are examples of dual use facilities.

Legacy of the Railroads

The most noteworthy American examples of the PPP model are the transcontinental railroads. Five of the six transcontinental projects between 1870 and 1900 were public–private partnerships. Edward Howard (1877) observed that "Among the social forces of the modern world the railroad holds unquestionably the first place. There is not a single occupation or interest which it has not radically affected. Agriculture, manufacturing, commerce, city and country life, banking, finance, law, and even government itself, have all felt its influence." The PPP relationships that financed the railroad agreements helped to promote the idea of American's manifest destiny in western territories by bringing political legitimacy and power to undeveloped areas. The new railroad infrastructure accelerated the increased access of goods to market and of consumer access to new manufactured and imported goods.

The partnership benefitted the railroad stockholders by deeding ten square miles of land for each linear mile of track constructed along the transcontinental routes. This enabled the private partners to sell the land or to leverage the real estate asset as collateral for loans in order to raise more working capital. One of the benefits of having the government as partner was the practice of carte blanche eminent domain which allowed the railroad to condemn and acquire lands owned by others in their path.

The railroad business model made money from multiple revenue sources. They not only generated income from freight and passenger fees, but they also took full advantage of the sale/lease and financial leveraging of the vast acreage of adjacent land parcels they acquired. The income from passengers and freight was from traffic traveling in both directions. This matches the business model of reusable rockets designed to carry people and cargo to and from space. Round-trip rockets are no trivial financial matter. Selling cargo capacity in both directions, to and from the surface of Earth, effectively doubles the ability to earn income from rocket payloads. Charles Perrow (2002), writing on the economic impact of the PPP model of the transcontinental railroads, observed that, "The railroad organizations ... were the biggest and richest organizations in the country, and then the world." Rockt vendors, take note.

The "Public" partners of the transcontinental railroad also benefitted. The Government received discount pricing for over 70 years (1870–1940) for such critical public expenses as mail contracts, moving government freight, passage for government personnel, and the transport of military equipment and troops. The partnership afforded no-bid government contract preferences (sole source transport), government endorsement and guarantee of Rail Road bonds (Bonds secured by land granted as loan collateral) that were often offered at above-market rates (e.g., 8%), financial benefits that included tax, tariff, license fee relief given to the railroad partner, and even third-party private equity investments: in other words, investment side-bets similar to "synthetic swaps" that were guaranteed by the US Government. Not a bad deal all the way around.

Possible benefits offered in a PPP agreement for a space project might include such elements as preferential payload contracts, tax and tariff breaks, R&D subsidies, dual use of launch and terminal assets, expansion of IP protections, and "relaxation" of anti-monopoly regulations. There is also the benefit of being an implicit infrastructure asset for military/intelligence activities, as well.

Some current examples of large-scale PPPs in the USA include the National Park Service and their ability to grant exclusive contracts for concessions such as hotels, food, in-park transportation, and other support services. Sports Stadiums in large metropolitan markets that are typically financed by local bond initiatives. The airline industry got its start with guaranteed income from Postal contracts, along with fare and route protection prior to deregulation. The telephone industry that was a protected monopolistic enterprise (ATT) so that mass long distance communications was readily available for commercial use. The model of the Antarctic science stations demonstrate the thinking behind the Space Treaty's prohibition from land ownership in space: Nations are prohibited from owning land in Antarctica but are free to conduct scientific research at will and "take" the scientific knowledge they've acquired during the partnership tenure.

SOE Business Model

Similar but not identical with the PPP model is the state-owned enterprise model (SOE) notably employed by the Chinese state. These preferred entities enjoy low-interest government loans and other preferential treatment that private competitor companies do not have.

The SOE is much like the PPP in the initial stages of development. However, there are key structural differences of funding, control, oversight and governance. In China, SOE funding is likely to remain permanently intact. In the West a PPP is dependent upon annual budget review and refunding in order to remain solvent. Officers of a SOE organization are often Chinese state bureaucrats or members of the Communist Party while officers in the PPP have an independent role and are selected from the private partner side. Finally, the SOE is protected as a virtual monopoly within the state structure and gets special preferential treatment because of its distinct state-related status causing an implicit competitive advantage that is based more on political influence and less on the quality of the goods or services provided by the SOE. A SOE has special preferential bargaining power with suppliers and unspoken pull with customers because of its close relationship with the central government. This is in contrast with a PPP enterprise which must compete in the open free market economy in order to maintain a legitimate and level playing field.

Vertical Integration: Controlling all the Key Elements of the Value Chain

One business model that has long been attractive to "old school" industrialists is vertical integration. Vertical integration simply means that the entire supply chain of raw materials and component parts is owned or tightly controlled by the parent firm. Often vertical integration extends beyond the production process and migrates to control of key distribution channels as well. This was the case for the motion picture industry until the U.S. Supreme Court sustained a lower court ruling that their monopolistic practices constrained open trade. An early notion of vertical integration comes from the frontier family farm business model where everything that was needed to produce the main crop was created on site.

Vertical integration is epitomized by Henry Ford's business model for manufacturing automobiles. His company controlled all the inputs to the production system. This included mining the coal to make the steel, owning the rail line and rolling stock that transported the steel to the factories, and owning the factories that turned the steel into the car frame and so on. He even dictated the dimensions of the crates used to ship parts to his assembly plants because he re-used the wood planks from the shipping crates as floorboards for his earlier car models. The problem with vertical integration is that the more production inputs are captured by the firm, the less flexibility the firm has in order to control production costs. Without competitive bidding for new suppliers, the overhead costs tend to increase which is reflected in a higher price charged to customers. As prices rise the company may become vulnerable to low-price competitors who are able to reduce their costs by using the open source business model (below) to acquire cheaper parts and components.

An offshoot of vertical integration is the company store, typified by rural Appalachian mining operations in nineteenth and early twentieth centuries.

Workers in the mine (or logging operation, or plantation, etc.) were housed in company housing, bought their food from the company store, educated their children in the company school, banked at the company bank. Every transaction was paid for with deductions from the worker's salary. Because goods and services offered were available only at the monopolistic company-owned store, workers had to pay artificially high prices, often on credit, and soon were irrevocably indebted to the company and became, in time, virtual indentured labor. This usurious business practice, while no longer legal in the USA, is a potential prototype for space-based operations in the far reaches of the solar system where workers would have limited access to alternative sources of goods and services other than the company they work for.

Vertical integration may be less of a strategic decision and more of a tactical result of the limits of operating in an immature economy like space. If there aren't a choice of suppliers handy to provide a company with much needed materials and services, then the firm may have no choice but to adopt a vertical integration business model, especially at first. At some future point the company may reevaluate the operations process and determine that the working environment is more mature, more competitive, more robust, and offers external resources that will allow it to switch from a vertical integration business model to an open source business model. Thus, reducing the burden of carrying the extra costs up and down the company's value chain.

While vertical integration looks good on a corporate planning white board there are many problems that result from having captured suppliers and distributors—among them the potential removal of a sense of competitiveness for better design, better quality, more efficient cost structures, effective customer service, and other legacy-related costs. A transition from vertical integration to an open business model will probably require a complete reorganization including a redefinition of the corporate culture. The antidote for ridged vertical integration is the open source business model.

Open Source Business Model (Outsourcing and Strategic Relationships)

Consider the open source business model in contrast to the closed model of vertical integration. The open source model acknowledges that it may be best to rely upon other firms to provide goods and services in a cost-efficient and competitive manner. This model outsources production of components to suppliers and contracts with independent distribution channels to get product to customers, thus minimizing overhead expenses. One drawback of operating in the space economy in any early phase of adoption rests on the question of available third-party suppliers who can deliver or distribute whatever materials are needed for the open source firm to meet performance expectations.

The open source business model considers what is vital to the firm—what needs to be controlled by the firm—and "buys out" what is not a core competency. Taco Bell doesn't raise its own beef or make its own tortillas, but instead uses a competitive bidding process to maintain quality standards at the best price they can negotiate. Likewise, Best Buy doesn't make the cameras or TVs it sells. And today even Ford uses external resources for non-core, non-critical activities. NASA doesn't build satellites or space probes; it buys out these elements from specialized vendors.

Other examples include such business services as ADT (payroll services), cloud data services, and office cleaning which a firm may purchase from an external provider. For very complex production challenges, like aerospace, key components are specified and then contracted with other manufacturers. In this way such sophisticated technologies as aircraft cockpits, landing gear, avionics, cabin interior, and jet engines are not necessarily made by the likes of Boeing or others but are farmed out to strategic partners in their supply chain.

For firms doing business in Space it seems likely that transportation, cargo and passenger brokers, life-support and habitat elements, communications technology, production machinery, and other capital equipment and professional services will be purchased from firms who will provide this mission-critical infrastructure. This is the short-term opportunity of the new frontier economy; anticipating the needs of firms who wish to participate in the new cosmos economy and providing those goods and services.

Long Tail Business Model

When plotting an entrepreneurial venture, the intuitive thing to do—the conventionally wise thing to do—is to aim for the heart of the demand curve. Because that's where most customers reside. Many business models assume that profitability is based solely on volume: The more customers you reach, the greater the sales, and the greater the sales, the higher the revenue. But that's not always so.

Figure 27.1 compares how most firms view their potential market (the middle high point of the distribution where most customers reside) versus the counterintuitive long tail (at the far right of the distribution where very few customers are). But the middle is where every competitor also wants to be,

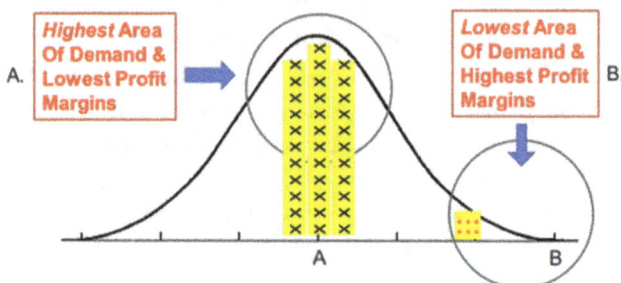

Fig. 27.1 Think of the area under the curve as representing the potential volume of demand for your product. The obvious place to be is where demand is clearly greatest; directly in the middle of the distribution curve (location A). The thin limited demand off to the right—the long tail of the classic "bell" curve—represents a unique pricing opportunity (location B). This is because the firm can premium price (charge a LOT more for the product). This high-margin solution circumvents the low-margin problem of targeting the middle of the distribution curve where other competitors congregate

which is why it is crowded with competitors who are forced to compete on price instead of service or quality, thus dramatically driving down profit margins for everyone. The potential customer base under the long-tail curve may be much less but the customers found at this end of the curve are willing to pay a premium for such products as rockets, oil tankers, fighter jets, space tourism, and customized cancer medications which can bring high prices and very healthy profit margins (Elberse 2008).

The emerging frontier economy of space is an excellent stage for the long-tail business strategy. As the new cosmos economy gradually matures in the foreseeable future there will be little infrastructure to help businesses survive and grow. Entrepreneurs who can provide specialty goods or services (and in space practically everything is a special good or service) will be able to exploit the low-volume/high-margin niche of the long-tail business model. Some of the businesses that will fall into this category include micro-gravity manufacturing equipment, transportation and shipping, cargo aggregation and freight forwarding, rockets and other space vehicles, habitats and life supporting equipment, high-tech space-based agri-business, specialized consulting services, rare disease medical treatments, and just about everything else during the early frontier phase of development.

Traditional Exchange Model

The exchange model is what most consumers are familiar with. When you go online, pull into your gas station, go shopping for groceries or clothing, visit your chiropractor, or just pay your utility bills you are completing a trade transaction. This trade of money for a service or for merchandise is the exchange business model.

Howard Bloom writing in *Scientific American* (2017), suggests the likely business opportunities in space will be mundane and analogous to life on the Interstate. A different kind of interstate, to be sure. His vision is a straightforward transposition of our view of commerce today. He suggests that commercial activities will consist of such businesses as "Gas stations" (propellant depots) that will function like turnpike rest stops. These space stations will act as commercial hubs of activity that spring up in the middle of nowhere on an endless stretch of freeway today. They will serve as "truck stops" and freight yards—logistics bases with cargo-handling equipment, centers of commerce providing whatever is needed to space tourists and industrial entrepreneurs on their way to the next place in space. Some waystations will be in cis-space, others may be on the surface of a celestial body. A few of these space-based strip malls will likely evolve to be the beginnings of space towns.

Traditional business enterprises that will use the exchange business model will include space tourism, retail, shipping, manufacturing, communications, and any other retail-like encounter, but recast with the innovation that comes with new technologies and IP licensing components and a strong dash of A.I. and robotic technologies.

Products that could find an immediate market would include vehicles like Earth-bound trucks, SUVs, and dune buggies—Moon-and-Mars ground vehicles, plus space tugs to haul loads in space. Fuel production equipment—units to turn the abundant water available on the Moon and Mars into rocket fuel, breathable oxygen, and drinkable water. Units to turn the carbon dioxide in Mars' atmosphere into plastics, graphene, and carbon fiber with which 3D printers can build more habitats, tools, and rovers along with units to turn the rusty rocks lying around on the Martian surface into high strength steel for habitats.

The Limits of Imitative Growth

Noted author Brink Lindsey, writing on "Frontier Economics" in *The American Interest* (2011), observed that, "the key distinction [of economic expansion] is between growth as more of the same and growth as something new and different." The danger of relying upon more of the same is that both markets and production capacity soon reach their peak where demand diminishes and production economies plateau. Lindsey noted that, "The former [growth by doing more of the same], or *imitative* growth, occurs within the existing technological frontier; the latter [growth as something new and different], or *innovative* growth, pushes that frontier outward."

Imitative growth depends on refining existing process and cutting costs. At some point the trend lines of this process (cost/benefit) approach an asymptotic limit and the benefits of pursuing this course of action become minimal. There is a way to break out of the convenient trap of a rinse-and-repeat autopilot business model operation: growth via innovation is the third and most powerful stage of the rocket. Adoption of the cosmos economy will enable economic growth as something new and different.

References

Bloom, H. (2017, January 4). *NASA should build a superhighway in space*. Scientific American Blog Network. Retrieved from https://blogs.scientificamerican.com/guest-blog/nasa-should-build-a-superhighway-in-space/

Choudary, S. (2013). Why business models fail—pipes vs. *Platforms*. Retrieved from https://www.wired.com/insights/2013/10/why-business-models-fail-pipes-vs-platforms/

Elberse, A. (2008, July 1). *Should you invest in the long tail?* Harvard Business Review. Retrieved from https://hbr.org/2008/07/should-you-invest-in-the-long-tail

Howard, E. (1877). *A railroad study*. Harpers New Monthly Magazine.

Lindsey, B. (2011, September 28). *Frontier economics*. The American Interest. Retrieved from https://www.the-american-interest.com/2011/09/28/frontier-economics/

Magretta, J. (2014, August). Why business models matter. *Harvard Business Review*. Retrieved from https://hbr.org/2002/05/why-business-models-matter

Ovans, A. (2015, January 23). What is a business model? *Harvard Business Review*. Retrieved from https://hbr.org/2015/01/what-is-a-business-model

Perrow, C. (2002). *Organizing America wealth, power, and the origins of corporate capitalism*. Baltimore, MD: Princeton University Press.

28

Diffusion of Innovations: The Five-Phase Adoption Model

A robust space-based market economy will not suddenly emerge fully formed from thin space. Different ventures will follow their growth curve at different rates. Regardless of the speed of adoption, space enterprises and the new markets they serve, will progress through five phases of adoption that are characterized by the operating environment of their enterprise, the constraints and advantages of their business model, and the readiness of customer demand. Building upon Everett Rogers' pioneering work, *Diffusions of Innovation* (1962), the five phases of space economy adoption are (1) Innovators, (2) Early Adopters, (3) Early Mainstream, (4) Late Mainstream, and (5) Lagging (Late) Adopters. These five phase categories best characterize the likely development and adoption of the new space economy.

Table 28.1 below is an integrated presentation of the five phases of adoption. This illustration is meant to set the stage for more detailed descriptions of each phase of the adoption model to follow.

Pioneering efforts to establish enterprises in the frontier phase will enable later phases to develop. A loose comparison with this future arena is Earth's current (2020s) modern global economy where nations in various stages of development interact with one another for mutual benefit. These are not just economic exchanges. It is hoped that a robust space trade will promote a similar exchange of goods, ideas, and relationships.

The five adoption phases presented below are prototypes. There are exceptions to their characteristics as well as likely overlaps in the transition from one phase to another. While these phases of adoption are presented as a linear sequence it is assumed that growth will be non-linear. Different micro economies in space will emerge and develop at different rates, progressing through

Table 28.1 The five phases of adoption are shown with approximate dates, likely business models that best fit each phase, and industries most likely to prosper in a given phase

Expanded Overview of the Adoption Model

	Phase 1	Phase 2	Phase 3	Phase 4	Phase 5
Phase Name	Innovators Frontier 2030 - 2060	Early Adopter 2050 - 2075	Early Mainstream 2070 - 2095	Late Mainstream 2090 - 2115	Late Adopter 2100 →
Phase Characteristics	No infrastructure, no consumer base, no industrial capacity, high investment potential, high entry/exit barriers	Significant economic activity causes tipping point from tentative nativity phase to solid growth phase	Broad base of investors. Initial wave of consolidation results in more stability, moderate entry barriers	Acceptance of sector as a) normal, b) integrated with Earth economy, c) viable, and d) robust & profitable	Economic autonomy, M&A, Space-based firms, P2P trade, very low entry/exit barriers
Likely Business Models	PPP, SOE, Private Capital	PPP, SOE, Open Business Model, VC investors	Open Business Model, Cartel investors, VCs, Long Tail Model PPP fades	Traditional Business model, Long Tail Model	Traditional Business model
Likely Industries & Markets	Communications Services, Infrastructure Systems, Robotics & AI, Transport to/from Earth	Human Habitat & Life Services, Energy Production & Distribution, Construction, Miro-gravity Manufacturing, Mineral Extraction & Processing, in-space transport services	Mass Food Production & Distribution, Services (Med, Legal, Security), Shipping/ Depot/ Distribution, Intra-Space transport	Manufacturing - Traditional, Tourism & Recreation, Transport - Planet surface	Consumer Products/ Services, Inter/intra space commerce, Trade Expansion
Likely Firms, Orgs., Sovereigns	Blue Origin, Orbital ATK, SpaceX, Boston Dynamics, VT MÄK, NASA, JAXA, China, Russia, UAE	Bigelow, BP, Exxon, Aramco, Bechtel, Fluor, Pharma - Glaxo, Gilead, Rio Tinto Mining, Pilot Co. transportation service centers	Bayer AG (Monsanto), Halliburton, P&G, Kraft Heinz, FedEx Logistics, Mersk, Cosco	Mitsubishi, Foxconn, GE, Marriott, Virgin, Tesla, Toyota	Blank Intentionally

different adoption phases, employing different business models across the solar system at the same time. This may seem chaotic and unorganized. Is it! The economic development of the solar system will not follow a predictable script. Such is the unpredictable nature of life in the future. This five-phase model of space economic development is based upon the research of Dr. Everett Rogers (1962) who pioneered the study of how innovation is adopted. Since his research on the phases of adoption was published it has been applied to various settings with predictable reliability.

Each phase description is meant to paint a broad archetypical portrait of the economic context so that those who wish to engage with the new space economy will have a better notion of how the context of their venture or investment and will affect the venture's success. The following descriptions of each phase are meant to be a general characterization and not a precise prescription. According to Rogers every new development process follows a somewhat predictable adoption curve. His model describes how a new idea, innovative product, or even a space-based economic sector will gain acceptance and adoption along a somewhat predictable S-Curve path.

The earliest phase of adoption, the Innovator phase, is the most vulnerable and a venture may encounter some rejection at first and even stagnate or die. If the initial hurdle of user acceptance is overcome, then innovators and risk-takers will begin to adopt the new venture as worthwhile. Rogers argues that adoption of this first phase will only reach about 2.5% of the full adoption potential. As information is circulated about the new opportunity and the benefits and risks become known in the marketplace, resistance to adoption decreases and investor acceptance and adoption increases.

Innovations like the telephone, commercial air travel, television set sales, the Hula Hoop, home PCs, iPhones, and Tesla automobiles followed a similar curve of adoption. Early adopters of these innovations tended to be people who were willing to pay premium prices for leading edge products or services.

The second phase, the Early Adopter phase, builds on the momentum of adoption that was kick-started by the first phase. The user adoption rate for this phase is estimated to increase at about three times that of the first phase to an additional 13.5% bringing the cumulative user acceptance to 16% by the end of the second phase. The third phase, called Early Mainstream, adds an additional 34% brings the cumulative adoption of the three early phases to only 50%.

The final two phases, Late Mainstream and Lagging Adapter, add 34% and 16%, respectively, bringing total user adoption potential to 100% as represented in the following chart.

Leaders and Laggards

Figure 28.1 revisits Fig. 23.1 discussed earlier but addresses and summarizes the different adoption rates of the five phases to demonstrate how adoption changes over time from the early innovator phase to the final lagging adapter phase.

When describing the characteristics of his model Rogers was careful to caution future researchers that his descriptions were based upon his observations only and that there were certain to be exceptions. He was concerned that some practitioners may take elements such as the rate of adoption, or even his descriptions of a typical adopter for a specific phase as a hard and fast rule instead of a broadly painted picture of an archetypical model which is what he said that he intended. He called them "ideal types" and observed that, "If no exceptions or deviations existed, ideal types would not be necessary. Ideal types are based on abstractions from empirical investigations." (p. 282).

The following discussion focuses on each of the individual five phases of adoption in the context of the new space economy.

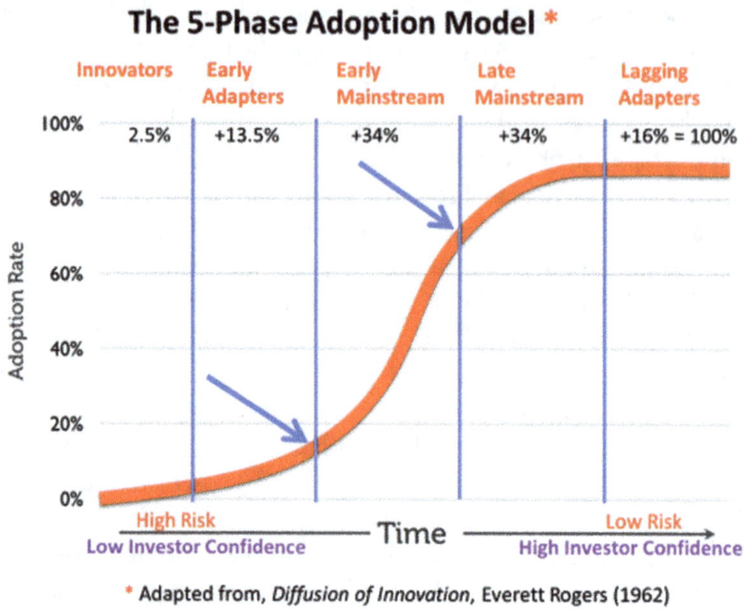

The 5-Phase Adoption Model *

Innovators	Early Adapters	Early Mainstream	Late Mainstream	Lagging Adapters
2.5%	+13.5%	+34%	+34%	+16% = 100%

High Risk / Low Investor Confidence — Time — Low Risk / High Investor Confidence

* Adapted from, *Diffusion of Innovation*, Everett Rogers (1962)

Fig. 28.1 The five phases are not evenly adopted. The first three phases, representing higher risk, are equal to the last two phases when there is higher confidence. The steepest growth rate occurs in phase 3 between the two inflection points represented by the arrows on the figure

Phase 1: Innovators/The Frontier Phase of the New Space Economy

The earliest commercial ventures in space will lay the foundation for later economic development for such activities as communications services, habitat construction, technical support systems, robotics and AI, transport to/from Earth, and other infrastructure-related investments.

Influencers, Innovators, and Risk-Takers

Everett Rogers used the word "venturesome" to describe those who engage at the very early phase of adoption. He went on to state that being venturesome, "… is almost an obsession with innovators," and that, "Their interest in new [opportunities] leads them out of a circle of local peer networks and into more cosmopolite social [and commercial] relationships." The term often used today to describe individuals who fit this profile is *influencer*—a trend-setter

28 Diffusion of Innovations: The Five-Phase Adoption Model

who defines the social and economic agenda. Rogers went on to offer a short list of traits vital to the entrepreneur who seeks to be an early entrant and adopt a new market opportunity. These include:

- Ability to manage substantial financial resources
- Ability to absorb potential losses form unsuccessful ventures (both financial loss and personal loss of ego)
- Ability to understand and apply complex technical knowledge
- Ability to cope with a high degree of uncertainty
- Ability to accept and deal with occasional setbacks
- Comfort operating in an environment of high risk

Rogers' list tends to focus more on humanistic capabilities or so-called soft skills. This is the skillsets of people who are shown to have a high emotional intelligence (EQ)), experience and maturity. (Goleman, 1995).

He further found that forward thinking innovators may not be respected by their peers, or those who aspire to be a peer. The innovator is, by nature, an outsider. They play a critical role by recognizing and acting upon new approaches to business solutions and by being a contributor to the flow of new ideas into a commercial system. By acting outside the mainstream, they bring a different perspective and help to define the nature of the new system and create a brand as a leader in the new arena. Elon Musk refused to accept the prevailing view that rockets were expendable and that the cost of payloads could be reduced, thus opening the door for his firm, SpaceX, to compete with the long-standing aerospace prime contractors.

Table 28.2 ties several key variables of this phase of space commerce together. In addition to a brief listing of phase characteristics, I've nominated several business models (previously discussed) that best fit this initial phase of adoption. Additionally, I suggest several likely industries and markets along with names of current companies who would probably do well to enter this

Table 28.2 Phase 1 Synopsis. Please note that all references to companies, organizations, and sovereign nations are for purposes of illustration only

	Phase Name	Phase Characteristics	Likely Business Models	Likely Industries & Markets	Likely Firms, Orgs., Sovereigns
Phase 1	Innovators Frontier 2030 - 2060	No infrastructure, no consumer base, no industrial capacity, high investment potential, high entry/exit barriers	PPP, SOE, Private Capital	Communications Services, Infrastructure Systems, Robotics & AI, Transport to/from Earth	Blue Origin, Orbital ATK, SpaceX, VT MÄK, NASA, JAXA, China, Russia, UAE

phase of development. I will repeat this broad analysis process as we progress through the next four phases.

Note the names of organizations and countries listed in column "E" are representative suggestions. It is yet unknown if the organizations mentioned here will play an active role developing or participating in the new space-based economy. These organizations were chosen because they are easily recognizable brands which helps to paint a picture of commercial space activity during this phase of adoption. This primary phase of adoption will attract a mix of corporations and nations. Sovereign nations will come to space to build their reputational capital and to establish a military foothold where feasible. Companies with a long-term agenda will see the new frontier of space as an opportunity to create a platform for future enterprise, to build infrastructure, to profit from mining operations, and to establish themselves as having a brand as a space innovator. Companies with a short-term expectation of return on their investment may enter the space market as a provider of habitats, tourist experiences, communications services, transportation-related equipment, and other business-to-business ventures.

But nothing is certain, it is just too soon to tell. (If you recall in Stanley Kubrick's film, *2001 a Space Odyssey*, there were images of Pan American Airways logos on the space station because that seemed to be a good guess at the time. Unfortunately, Pan Am is no longer taking reservations. Such are the vagaries of corporate mortality.)

This phase, the formative frontier stage, is best characterized by uncertainty and risk. Early entrants may seek interdependence with other early market players for synergies across industries, supply chains, and technologies. Unlike other Earth-bound frontier markets, space will neither provide readily exploitable labor nor local consumers for purchase of goods or services. In-situ resource utilization (ISRU), turning materials found in space into usable products instead of transporting finished goods from Earth, will rely upon specialized equipment like 3D printers to manufacture components for making structures and other necessary components. Because this early phase will require building a complete industrial platform from scratch the range of venture opportunities is tremendous. Providing infrastructure to later phases may be among the most profitable areas of early investment, setting the stage for long-term rents and fees as the market matures and additional new firms enter the market arena in later phases of adoption.

Those companies who establish themselves earliest will likely benefit most. Probable business models include public–private partnership (PPP), State owned enterprises (SOE)), and cartels of firms sharing risks and rewards. Likely industry sectors may include communications services, customized habitat, technology, infrastructure systems, engineering and construction, transport and trans-shipping logistics management systems, and robotics providers. Space Tourism will likely play a significant role and will evolve over time with more attention paid to customer services and amenities. Likely firms who may participate at this phase may include Blue Origin, Orbital ATK, SpaceX, Bechtel, Fluor, Boston Dynamics, and Bigelow. Sovereign nations such as China, India, Israel, U.S., and Japan will likely have an active presence.

The general direction of activities will originate from Earth and target extraterrestrial industrial customers. Commercial locations will be situated in orbit around Earth, parked in cislunar space between the Earth and the Moon, on the surface of the Moon, or eventually proximate to Mars as well as some asteroid locations, or other solar system bodies. The prospect of easily maneuverable space outposts located in space neither in permanent orbit nor anchored on the surface of a moon or planet, may come to be a practical alternative to permanent operations planted firmly planet-side. Instead of building new construction machinery at a new mining site it will be easier to bring a complete set of construction machinery to the next asteroid.

Innovators, influencers and entrepreneurs are often mislabeled as gamblers who take unwarranted chances and who blindly trust their fate to luck. This is only partly true. Just as a professional poker player doesn't trust the outcome of a game to luck but calculates the odds in her favor before placing a bet or folding her cards, so too do professional business innovators rely on extensive knowledge about market potential, competition, available suppliers, entry and exit barriers, and a host of other critical data before applying their intuition. Dr. Daniel Kahneman (2015), the Nobel Prize winning economist who is recognized for his research in behavioral economics and the psychology of judgment and decision-making, contends that the best decisions are made with a healthy mix of objective and subjective considerations. Intuition tempered with data. That is the secret sauce of successful politicians and serial entrepreneurs. At the end of the first World War the Treaty of Peace (aka: The Treaty of Versailles) imposed severe sanctions on the Germany people by emotionally charged generals and politicians. As a result, Germany's economy was unable to rebuild, the society became economically and politically unstable, which resulted in the rise of Facism and Hitler. In contrast, the more logical and even-handed Marshall Plan at the end of World War II created a way for Germany to rebuild itself and rejoin the world community. The emotionally charged Treaty of Versailles caused more damage instead of healing a wounded world, while the consideration of the Marshall Plan helped to stabilize the global balance of power.

Figure 28.2 depicts the flow of goods, services, and investment capital for phase 1 of the adoption model. Note the one-way direction of transactions as represented by the arrows. This is meant to indicate Earth's dominant commercial and political role as it influences the control and distribution of capital and goods during this period of industrial space development. Destinations X, Y, and Z indicate various hypothetical settlement and commercial outposts emerging in this early phase such as cislunar, asteroids, planets, or deep space locations. Phase 1 will yield approximately only 2.5% of the potential market adoption.

Phase 2: Early Adopter Phase

Rogers characterized the general profile of phase 2 of early adopters as dependent upon phase 1 adopters for information and partnerships. In this phase early adopters, "… help trigger the critical mass," of the entire adoption process. Early adopters are still far ahead of the general commercial investor and will likely set the stage for how fast and robust commercial adoption will occur. Rogers suggested that early adopters may be viewed as "missionaries" because they tend to be well respected by their peers and they provide feedback to the commercial community about the efficacy of the adoption

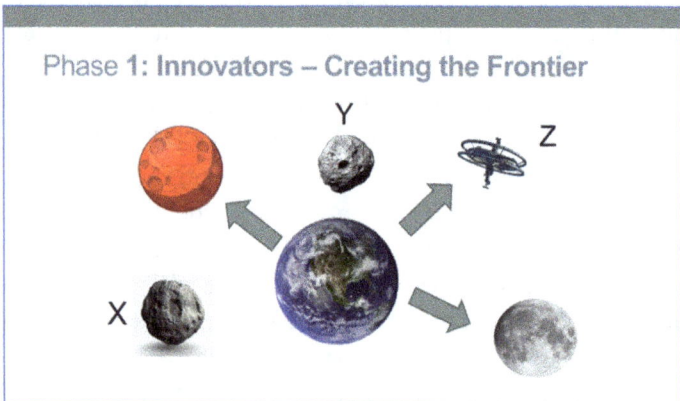

Fig. 28.2 Phase 1: Innovator Phase (Frontier Phase). Earth is central hub of Frontier/Innovator Market Economy. Economic activity is defined and controlled by Earth-based activities such as commercial investment, technology, and rudimentary mining and space-based manufacturing. Destinations X, Y, and Z indicate the possibility of various industrial outposts, settlements and commercial activities including cislunar, asteroids, planets, or deep space locations. This phase will see investments in primary infrastructure. Space policy agreements will begin to be challenged as industrial momentum increases

process. Early adopters play an active role as evangelists/salespersons extolling the benefits of the innovation or market. The use of satellite data to aid agriculture production wasn't initially accepted but early adopters of GPS demonstrated the benefits of accuracy and cost savings. Now it is common to see automated tractors harvest grain guided by signals from space.

Table 28.3 presents the second phase, a transition or tipping point from a frontier economy to a growth economy. This phase is characterized by the increase of two-way commercial transactions typified by the beginning of in-space processing of minerals mined from asteroids, the construction of habitats and outposts, the production of pharmaceuticals, and the development of large-scale logistics operations including transportation and warehousing. This transition phase sets the stage for the third phase; growth and expansion.

Just as with our discussion of phase 1 please note that this is meant to represent an *ideal* model to create a general understanding of the nature of this phase. Again, the organizations mentioned in column "E" are meant to be representative only.

In this second phase, early commercial activities will likely coalesce to take advantages of natural synergies that come of operating in the unpredictable space setting. Lessons learned from early entrants in phase 1 will benefit those firms who follow who will provide a range of business services and supplies to new ventures and outpost communities. Keiretsu-like partnerships that are built of both vertical and horizontal relationships may be necessitated by

Table 28.3 Phase 2 Synopsis. This table shows the relationship between phase 1 and phase 2. Phase 2 is strongly dependent upon the degree of success of the first phase in establishing a successful foundation. With a broad spectrum of different business sectors established in phase 2 comes a mix of business models and organizations eager to enter and adopt the new environment

	Phase Name	Phase Characteristics	Likely Business Models	Likely Industries & Markets	Likely Firms, Orgs., Sovereigns
Phase 2	Early Adopter 2050 - 2075	Significant economic activity causes tipping point from tentative nativity phase to solid growth phase	PPP, SOE, Open Business Model, VC investors	Human Habitat & Life Services, Energy Production & Distribution, Construction, Miro-gravity Manufacturing, Mineral Extraction & Processing, in-space transportation services	Bigelow, BP, Exxon, Aramco, Bechtel, Fluor, Pharma - Glaxo, Gilead, Rio Tinto Mining, Pilot Co. transportation service centers
Phase 1	Innovators Frontier 2030 - 2060	No infrastructure, no consumer base, no industrial capacity, high investment potential, high entry/exit barriers	PPP, SOE, Private Capital	Communications Services, Infrastructure Systems, Robotics & AI, Transport to/from Earth	Blue Origin, Orbital ATK, SpaceX, VT MÄK, NASA, JAXA, China, Russia, UAE

operating in an immature and physically unforgiving economic setting. In this approach companies may join and agree to own a joint stock ownership position in one another's companies to help mitigate risk and enable a longer-term strategic planning horizon. Those who choose to establish partnerships with first movers from phase 1 will find mutual benefit.

As more firms establish themselves in the new space economy issues of dominance and control will probably emerge. Those who are underfunded may scale back, retreat, or merge with a competitor. Business models that successfully enabled firms in phase 1 will compliment other business models such as the open model or the long-tail model (discussed previously) to gain traction. Additional sectors likely to succeed will include human habitat and life services, energy production and distribution, engineering and construction (both on planet surface and in cislunar space), micro-gravity manufacturing, mineral extraction and processing, and more sophisticated tourism experiences. It is also possible that early examples of warehousing, distribution and logistics will begin to expand at this phase. Likely firms may include Aramco, BP, Glaxo, Gilead, Rio Tinto, Pilot transportation service centers, and Royal Caribbean Cruises who will target the new space business and leisure travel customers.

A Fork in the Road

Figure 28.3 introduces a new element in the flow of goods, services, and investment capital. If the "down-mass problem"—the technical hurdle of shipping large volumes of freight from space to Earth has been resolved, then space-based enterprises will serve Earth-based customers. If the down-mass problem is not overcome, if shipping large commercial payloads to Earth remains a logistical roadblock, then products originating in space will target space-based commercial customers. Producers of goods in space will adjust their output to best serve their customers' needs. In this early phase of adoption individual consumers will be located on Earth and industrial customers will be in space for some time to come. Phase 2 will yield approximately only 1 of full potential.

Phase 3: Early Mainstream Phase

Phase 3, the early mainstream phase of growth and expansion will see increased economic activities beyond the original B2B industrial categories. Like the 1849 California Gold Rush, entrepreneurs will find unanticipated and

28 Diffusion of Innovations: The Five-Phase Adoption Model

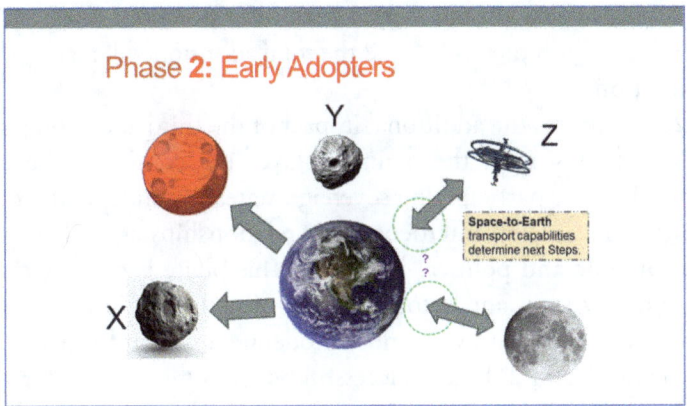

Fig. 28.3 Phase 2: Early Adopters. Resolution of the "down-mass" problem will determine direction and focus of this phase of development. If down-mass technologies are developed, then Earth-based demand will drive the space economy. If this problem is not resolved, then space-based demand will define the cosmos economy. Space-driven economy will set a tone of greater independence from Earth and focus growth on other markets in the solar system. Frontier to Early Adopter transition. Trade is still highly dependent upon Earth as a market partner. Note: the "?" indicates the unresolved question about the commercial ability to ship large payloads of goods from space producers to Earth markets: The down-mass problem

original opportunities to provide new goods and supporting services to customers located in space. Investment in space-based enterprises will continue to rise. There will be increased adoption of this new emerging market's viability. This phase will herald the closing of frontier phase activities located closer to Earth but will initiate the expansion of new frontiers located beyond Mars, near the asteroid belt, and into the farther reaches of the solar system where unexploited resources and new commercial opportunities may be found. New economic trading hubs and centers will emerge to fill demand in these new frontiers.

The focus of the first two phases is to introduce and then advance the overall adoption of space commerce. This is a slow and deliberate process as the adoption process takes hold in the broader economic community. The move to the third phase of adoption, the Early Mainstream phase, signals a more accelerated and widespread adoption. Consider the S-curve progression of earlier Fig. 28.1, the third phase represents the steep growth portion of the curve that links the beginning of the adoption process with the rest of the model. In describing the significance of this phase Rogers quotes Alexander Pope from *An Essay on Criticism* (1711, 1917), "Be not the first by whom the new is tried, nor the last to lay the old aside."

Phase 3 will yield approximately 34% of the potential market adoption. When combined with phases 1 and 2 the total adoption will reach 50% of full market adoption.

Table 28.4 captures the additional impact of the third adoption phase. This phase is the driving gear, the mainspring, of the coming space economic engine. In this phase participating sovereign states and independent commercial ventures will develop interdependent relationships as well as establish a sense of economic and political hierarchy. This phase is characterized by its broad base of investors and actors. A wave of consolidation may produce a more stable economic base with more moderate entry and exit barriers than earlier in phases 1 and 2. Likely successful business models will include more traditional Earth-based businesses as well as those already discussed.

It is safe to assume there will always be new frontier-level economic centers emerging in the solar system's economy. This will result from free market expansion as new investors and entrepreneurs push further and further out from Earth in search of new economic opportunities and markets to commercially exploit. Like the American western frontier expansion of the late nineteenth century, earlier commercial activities will incubate platforms for the continued creation of new frontier economic centers (i.e., outposts and

Table 28.4 Phase 3 Synopsis represents the early mainstream adoption. In addition to an expansion of industry sectors and markets this phase is noteworthy because of the development of intra-space trade as the major focus of industrial space economic expansion

	Phase Name	Phase Characteristics	Likely Business Models	Likely Industries & Markets	Likely Firms, Orgs., Sovereigns
Phase 3	Early Mainstream 2070 - 2095	Broad economic base of investors and actors. Initial wave of consolidation results in more stable economic base, moderate entry	Open Business Model, VC investors, Long Tail Model, PPP fades	Food Production & Distribution, Services (Med, Legal, Security), Shipping/ Depot/ Distribution,	Bayer AG (owns Monsanto), Halliburton, P&G, Kraft Heinz, FedEx Logistics, Mersk, Cosco
Phase 2	Early Adopter 2050 - 2075	Significant economic activity causes tipping point from tentative nativity phase to solid growth phase	PPP, SOE, Open Business Model, VC investors	Human Habitat & Life Services, Energy Production & Distribution, Construction, Miro-gravity Manufacturing, Mineral Extraction & Processing, in-space transportation services	Bigelow, BP, Exxon, Aramco, Bechtel, Fluor, Pharma - Glaxo, Gilead, Rio Tinto Mining, Pilot Co. transportation service centers
Phase 1	Innovators Frontier 2030 - 2060	No infrastructure, no consumer base, no industrial capacity, high investment potential, high entry/exit barriers	PPP, SOE, Private Capital	Communications Services, Infrastructure Systems, Robotics & AI, Transport to/from Earth	Blue Origin, Orbital ATK, SpaceX, VT MÄK, NASA, JAXA, China, Russia, UAE

28 Diffusion of Innovations: The Five-Phase Adoption Model

settlements) that are likely to develop into autonomous emerging markets and eventually grow to mature communities and urban centers. There will always be a frontier economy spawning new ventures somewhere in the solar system for the foreseeable future. New sectors adding to previously established commercial activities at this phase may include food production and distribution, services (i.e., medical, legal, tech, security, consulting), and transportation management both in space and planet-side. Examples of more firms may be typified by companies like Halliburton, Bayer (Monsanto), P&G, GE, Cosco, Maersk, and Cargill.

Figure 28.4 notes the anticipated increase in *intra*-space-based transactions as indicated by the arrows on the chart below. In this phase autonomous trade among sources and destinations exclusive of Earth will signal a major shift in thinking about the role of Earth regarding trade and finance in industrialized space. This shift will also impact the flow of capital away from Earth with more economic activities centered in space.

Phase 4: Late Mainstream Phase

Phase 4, the emerging market phase, is the transition from rapid expansion to stability and maturity. This may be misleading because as economic activities far above low Earth orbit (LEO)) reach a level of maturity, the further reaches of the solar system will simultaneously incubate an increasing array of

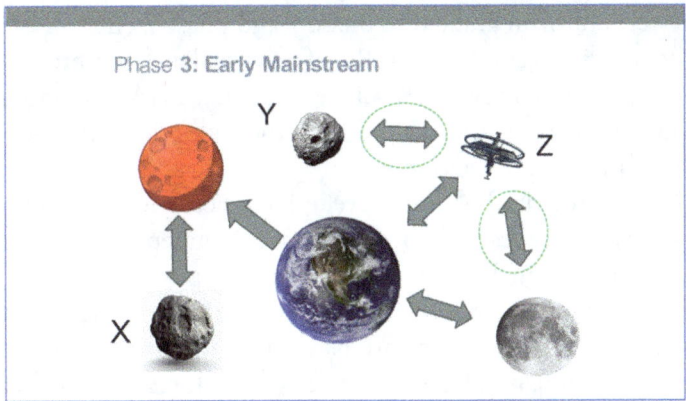

Fig. 28.4 Phase 3: Early Mainstream. Trade among established space centers indicates increasing independence from Earth. Imposition of regulations and standardized accountability metrics. New outposts, trade hubs, far-flung communities, and urban centers. Increase of consumer-related goods and services

brand-new frontier markets. This suggests that the progression from phase to phase is not always neatly linear when considering the economic potential of the entire solar system. While some areas of the new space economy will move more rapidly to become a mature economy, others will not progress as quickly and will remain longer at frontier or intermediate states. In the colonial 1700s Boston, New York, and Philadelphia were established sophisticated centers of commerce while at the same time Los Angeles was still a small rural agricultural village.

With the onset of the fourth phase Rogers estimates that adoption will have significantly surpassed the halfway mark. This penultimate phase signals the cusp of saturation and maturity; The level of competition has increased, the untapped opportunities that were the low-hanging fruit for those innovators and early adopters are scarce, and the shift in business models from the highly leverage Long-Tail to more traditional commercial trade transactions means that this phase represents a new era where highly competitive commodity pricing and steadily decreasing margins may result from an increasing consumer base.

Rogers maintains that late mainstream adopters have a "skeptical" approach to innovation and new ventures and, "… do not adopt until most of the others in their [market sector] have already done so." They are late bloomers who are risk-averse and are most comfortable playing it safe. They typically come into the game with limited resources which means, according to Rogers, "… that most of the uncertainty about a new [venture] must be removed before the late mainstream feel it is safe to adopt." Venture capital investors will likely reduce their time horizon regarding their expectations of return from the generous long-term horizon granted to phase 1 and phase 2 enterprises and then revert to demanding a shorter-term ROI for later phase entrepreneurs.

Phase 4 will yield another 34% of the potential market adoption. When combined with earlier phases 1, 2 and 3 the total adoption will reach approximately 84% of full market adoption.

Table 28.5 summarized phase 4 development activities and shows that the space economy is now capable of full economic independence from Earth. Lack of dependence on Earth indicates a critical power shift that will lead inevitably to potential economic autonomy for an increasing number of space-based enterprises and the various communities they serve. The value proposition for companies based in space will no longer demand referral to the uniqueness of the goods due to their production in space, but goods originating in space will be competitive simply because of the traditional reasons of quality, price, utility, or convenience.

28 Diffusion of Innovations: The Five-Phase Adoption Model

Table 28.5 Phase 4 Synopsis. With the coming of investors in phase 4, adoption is estimated to reach roughly 84% saturation. This heralds a mature market with sellers competing more on price in lieu of such intangibles as quality and service. Earth is no longer an integral component

	Phase Name	Phase Characteristics	Likely Business Models	Likely Industries & Markets	Likely Firms, Orgs., Sovereigns
Phase 4	Late Mainstream 2090 - 2115	Acceptance of sector as a) normal, b) integrated with Earth economy, c) viable, and d) robust & profitable	Traditional Business model, Long Tail Model	Manufacturing - Traditional, Tourism & Recreation, Transport - Planet surface	Mitsubishi, GE, Foxconn, Marriott, Virgin Galactic, Tesla, Toyota
Phase 3	Early Mainstream 2070 - 2095	Broad economic base of investors and actors. Initial wave of consolidation results in more stable economic base, moderate entry	Open Business Model, VC investors, Long Tail Model, PPP fades	Food Production & Distribution, Services (Med, Legal, Security), Shipping/ Depot/ Distribution,	Bayer AG (owns Monsanto), Halliburton, P&G, Kraft Heinz, FedEx Logistics, Mersk, Cosco
Phase 2	Early Adopter 2050 - 2075	Significant economic activity causes tipping point from tentative nativity phase to solid growth phase	PPP, SOE, Open Business Model, VC investors	Human Habitat & Life Services, Energy Production & Distribution, Construction, Miro-gravity Manufacturing, Mineral Extraction & Processing, in-space transportation services	Bigelow, BP, Exxon, Aramco, Bechtel, Fluor, Pharma - Glaxo, Gilead, Rio Tinto Mining, Pilot Co. transportation service centers
Phase 1	Innovators Frontier 2030 - 2060	No infrastructure, no consumer base, no industrial capacity, high investment potential, high entry/exit barriers	PPP, SOE, Private Capital	Communications Services, Infrastructure Systems, Robotics & AI, Transport to/from Earth	Blue Origin, Orbital ATK, SpaceX, VT MÄK, NASA, JAXA, China, Russia, UAE

This phase of development will be characterized by its acceptance as (a) competitive with Earth's economy, (b) neither a unique nor an overly risky platform for corporate investment, (c) an economic force of weight and legitimacy, and (d) the new normal. Additional sectors added to the mix are likely to include deep space tourism and recreation, durable goods manufacturing, agriculture on a larger industrial scale (both food and fiber), commercial and consumer electronics, and retail commercial ventures to serve the growing human population in far-flung space communities. Examples of additional firms who may participate are Foxconn (Hon Hai Precision Industry), Trump Organization, Motel 6 (Bigelow), Mitsubishi, Amazon, Toyota, Sanyo, Virgin Group, and Tesla are all likely representative of firms establishing space-centric brands in this phase.

Figure 28.5 presents the idea that space trade will have the capability to evolve and mature without requiring Earth at the center of industrialized space. The growth of the cosmos economy is now autonomous and is independent from Earth for political, military, or economic subsidy. The hegemony of Earth's role in determining financial, political, social, and military power and policy will be challenged as the extraterrestrial economy begins to operate under its own steam. Earth, essentially, is demoted from the central character to a subordinate role.

Phase 5: Late Adopters: Market Maturity Phase

The final phase of this adoption model, market maturity, will see the emergence of a self-sustaining economic system no longer dependent upon Earth for resources or economic direction that is politically interconnected and commercially integrated with an increasing number of commercial outposts and settlements. At this stage of development there will be a critical mass of producers and customers located off-planet to support a long-term sustainable and competitive advantage with Earth. The momentum from this growth will seek continued growth and expansion in an increasing inventory of non-Earth markets. Just as St. Louis, Missouri was the gateway city to the American Western frontier in the late 1800s, the jumping-off point for settlers and wagon trains seeking new opportunities in the undeveloped frontier west of the Mississippi River, the new space centers of commerce beyond LEO will enable the next wave of entrepreneurs, investors, and adventurers seeking prosperity and new market opportunities in the continually growing space economy.

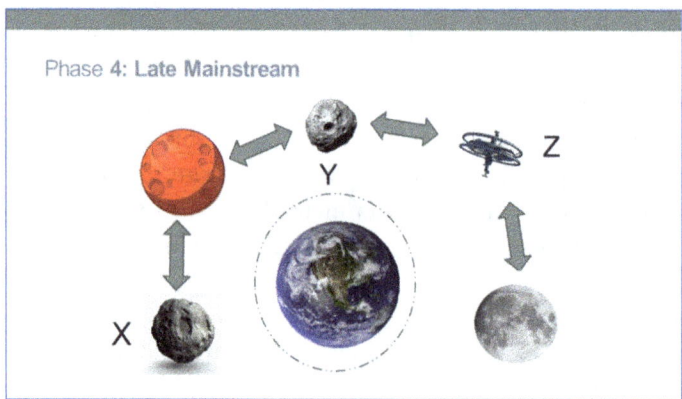

Fig. 28.5 Phase 4: Late Mainstream. Expansion to Mature Market transition. Trade is no longer reliant upon Earth-based markets (represented by dotted line encircling Earth). Greater participation of consumer-oriented manufacturing

Table 28.6 is the fully realized adoption model. This phase may be characterized by increasing M&A activities (corporate mergers and acquisitions), an increase of space-based financial institutions, more reliance on trade among non-Earth commercial centers, and much lower entry and exit barriers such as access to capital, robust supply chains, and doing business in an established market setting. More traditional business models will prevail along with business models employed earlier in the development cycle such as Long-tail, PPP, and SOE. Sectors in this phase will now include greater representation from providers of consumer goods and services. It is likely that a stock exchange representing off-planet ventures will attract investors to firms in a dynamic autonomous economy. [Note: trading on space-based exchanges may be problematic due to the long and inconsistent transmission lag times for buy or sell transactions from different locations in space. Stay tuned.]

Rogers labeled this late adoption phase as "Laggards." He recognized the potential pejorative connotation of the term describing those who finally decided to jump into the game at the very last phase of development. But he justified this term because, "… most non-laggards have a strong pro-innovation bias." Regardless of the term, he accurately makes the point that those who react slowly to opportunity will have little to make their own except to scramble to glean the leavings of those more courageous who took the risks in earlier phases.

Phase 5 may well signal the independence and maturity of non-Earth centers of commerce. These new economic hubs are likely to take on different shapes and operate under rules all their own. The significance of this development is that Earth's role will probably be relegated to that of a co-equal actor in the new industrialized space economy. The hallmark of this phase will be the flow of commerce between other commercial centers in space independent of Earth. At this singularity, products bearing the imprimatur of "Made in Space" will no longer be considered special or noteworthy—just business as usual.

Figure 28.6 depicts how former outposts and settlement communities may mature to become new central commercial hubs and urban centers. As these communities mature and grow it is likely they will spawn new frontier (phase 1 type) offshoots and initiate a repeat of the adoption process previously described.

Phase 5 will potentially yield the remaining 16% of the potential market adoption. When combined with earlier phases 1, 2, 3, and 4 the total adoption will reach approximately 100% of full market adoption.

The momentum to develop industrial space and the accompanying activities surrounding establishing permanent livable off-planet settlements is anticipated

Table 28.6 Phase 5 Synopsis. This is the full five-phase model. Dates indicated at the top of each column are approximations to suggest an overlapping sequence of progression. The nations, organizations, and firms are represented as likely examples only

The Complete Five Phases of Adoption

	Phase 1	Phase 2	Phase 3	Phase 4	Phase 5
Phase Name	Innovators Frontier 2030 - 2060	Early Adopter 2050 - 2075	Early Mainstream 2070 - 2095	Late Mainstream 2090 - 2115	Late Adopter 2100 →
Phase Characteristics	No infrastructure, no consumer base, no industrial capacity, high investment potential, high entry/exit barriers	Significant economic activity causes tipping point from tentative nativity phase to solid growth phase	Broad base of investors. Initial wave of consolidation results in more stability, moderate entry barriers	Acceptance of sector as a) normal, b) integrated with Earth economy, c) viable, and d) robust & profitable	Economic autonomy, M&A, Space-based firms, P2P trade, very low entry/exit barriers
Likely Business Models	PPP, SOE, Private Capital	PPP, SOE, Open Business Model, VC investors	Open Business Model, Cartel investors, VCs, Long Tail Model PPP fades	Traditional Business model, Long Tail Model	Traditional Business model
Likely Industries & Markets	Communications Services, Infrastructure Systems, Robotics & AI, Transport to/from Earth	Human Habitat & Life Services, Energy Production & Distribution, Construction, Miro-gravity Manufacturing, Mineral Extraction & Processing, in-space transport services	Mass Food Production & Distribution, Services (Med, Legal, Security), Shipping/ Depot/ Distribution, Intra-Space transport	Manufacturing - Traditional, Tourism & Recreation, Transport - Planet surface	Consumer Products/ Services, Inter/intra space commerce, Trade Expansion
Likely Firms, Orgs., Sovereigns	Blue Origin, Orbital ATK, SpaceX, Boston Dynamics, VT MÄK, NASA, JAXA, China, Russia, UAE	Bigelow, BP, Exxon, Aramco, Bechtel, Fluor, Pharma - Glaxo, Gilead, Rio Tinto Mining, Pilot Co. transportation service centers	Bayer AG (Monsanto), Halliburton, P&G, Kraft Heinz, FedEx Logistics, Mersk, Cosco	Mitsubishi, Foxconn, GE, Marriott, Virgin, Tesla, Toyota	Blank Intentionally

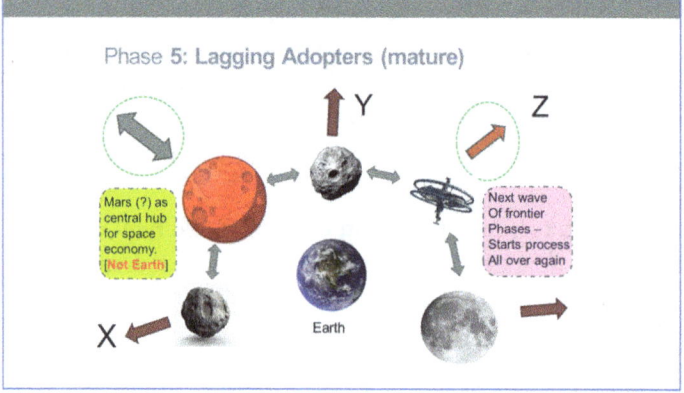

Fig. 28.6 Phase 5: Lagging Adopters (AKA: Maturity Phase). Mature/Independent Market Economy. Closed loop: Self-sustaining demand and supply completely in space. Establishment of new commercial hubs and communities. Growth will come from pushing farther out in the solar system, thus creating an endless succession of new frontiers

to be well underway or accelerate in earnest by about the mid-twenty-first century. Unknown is the level of participation in these efforts by sovereign states such as China, Russia, Japan, India, the EU, and, of course the United States. These political players will want to carve out significant influence in the new space frontier and therefore have input to its scope and momentum.

The costs of such a transnational (trans-planetary?) commitment are not trivial by any means. Even with PPP's carrying the bulk of the financial load of the early phases of commercial space investment opportunities—for such critical pieces as infrastructure, energy acquisition, human habitats, agriculture, and many other key components needed to support human settlements in space—there is only so much uncommitted investment capital (either public or private) to finance the startup of a whole new economic system. The opportunity cost of diverting funds and resources to space investments may be prohibitive; the long-term impact of these anticipated enterprises may well outweigh an investor's short-term anticipated return; A very risky business from an investor's point of view.

While there may not be a master plan that defines the scope and scale of the new space economy, there is a predictable path that can be described by the five overlapping phases of development outlined above. These five phases describe the path from a frontier economy, then to an emerging economy, and finally to a robust and mature independent economy. Each of the five phases may be best matched with appropriate business models so that an entrepreneur or investor may better judge when and how to enter the new off-planet economy. The long-term gains of investors and corporate ventures may well be incalculably astronomical (sic). First movers, the early private or state investors, will be best positioned to carve out the best pieces of the pie for themselves with a long-term expectation of lucrative rents and profits from their investment risks.

As more companies and independent investors seek participation in the new economy the economic platform will become more robust, will mature and stabilize, and will gain greater economic self-sustainability. The benchmark end state will be when all economic activities, from raw materials acquisition, to production, and finally to consumption, will be transacted in a closed loop that is independent of Earth, when space-based producers will sell to space-based customers.

References

Goleman, D. (1995). *Emotional intelligence*. New York: Bantam Books.

Kahneman, D. (2015). *Thinking, fast and slow*. New York: Farrar, Straus and Giroux.
Pope, A. (1917). *Pope essay on criticism*. Cambridge University Press. (Original work published 1711).
Rogers, E. M. (1962, 2003). *Diffusion of innovations*. Free Press. (Original work published 1962).

29

Connecting the Dots

Now we have a complete and comprehensive picture of how each of the five phases will work interactively to build one upon the next and create something new and significant: a space-based economic engine. It is important to pause and step back a bit from this very linear model. It is misleading. If this model of adoption is taken alone the greater implication can easily slip past us; it's a classic case of not seeing the entire forest because of the trees nearby. In fact, the multi-phase process of adoption will repeatedly replicate and take place all over the solar system at different rates of progression and with different degrees of success.

In short order there will be multiple frontiers attracting high-risk venture capital and high-concept innovators hoping to carve out new autonomous economic zones in space. Some of these will fail, others will thrive and move on to successive phases of adoption. Some will learn, or fail to learn, from those who have gone before. Consider the implications of multiple phase one activities feeding into multiple phase two, three, four, and five commercial centers. The progression of adoption doesn't have to be constrained to only one cluster of ventures but can reach across to multiple others feeding into and benefitting from all the other processes of adoption going on at the same time in the industrialized solar system.

A Reversal of Fortune

Given the size of our solar system and the estimated vast resources distributed within it, there will likely always be a frontier phase developing somewhere in space. The dire warnings of Frederick Jackson Turner at the Chicago World's Fair in 1893, about the crushing impact of the end of the frontier on civilization, can now be reversed. With an endless supply of frontiers for adventurers, explorers, entrepreneurs, and social outcasts to benefit, our human experience in space will continue to expand. We are at the blueprint stage of the cosmos economy. In the coming decades the organic growth of free markets in space will follow their own course to define how the space economy will grow and mature. In this dynamic context, economic centers of greater and lesser development will all exist together, each benefitting from their unique differences.

Table 29.1 combines the business model examples earlier discussed with the five phases of adoption that describe the steps the cosmos economy will likely traverse from infancy phase (Frontier Economy), through the Early Adopter phase (Expansion Economy), and finally to Lagging Adopter phase (Mature Economy). We can now construct a likely roadmap of development

Table 29.1 The complete adoption model synopsis. This figure shows the combined relationships between the five phases, likely business models, industries, and markets, along with the types of firms that will conduct business in various space settings. The dotted line S-Curve superimposed on the figure shows the cumulative degree of adoption from early frontier through late adopter phases

Connecting the Dots
The Complete Adoption Model

	Phase 1	Phase 2	Phase 3	Phase 4	Phase 5
Phase Name	Innovators Frontier 2030 - 2060	Early Adopter 2050 - 2075	Early Mainstream 2070 - 2095	Late Mainstream 2090 - 2115	Late Adopter 2100 →
Phase Characteristics	No infrastructure, no consumer base, no industrial capacity, high investment potential, high entry/exit barriers	Significant economic activity causes tipping point from tentative nativity phase to solid growth phase	Broad base of investors. Initial wave of consolidation results in more stability, moderate entry barriers	Acceptance of sector as a) normal, b) integrated with Earth economy, c) viable, and d) robust & profitable	Economic autonomy, M&A, Space-based firms, P2P trade, very low entry/exit barriers
Likely Business Models	PPP, SOE, Private Capital	PPP, SOE, Open Business Model, VC investors	Open Business Model, Cartel investors, VCs, Long Tail Model PPP fades	Traditional Business model, Long Tail Model	Traditional Business model
Likely Industries & Markets	Communications Services, Infrastructure Systems, Robotics & AI, Transport to/from Earth	Human Habitat & Life Services, Energy Production & Distribution, Construction, Miro-gravity Manufacturing, Mineral Extraction & Processing, in-space transport services	Mass Food Production & Distribution, Services (Med, Legal, Security), Shipping/ Depot/ Distribution, Intra-Space transport	Manufacturing - Traditional, Tourism & Recreation, Transport - Planet surface	Consumer Products/ Services, Inter/Intra space commerce, Trade Expansion
Likely Firms, Orgs., Sovereigns	Blue Origin, Orbital ATK, SpaceX, Boston Dynamics, VT MÄK, NASA, JAXA, China, Russia, UAE	Bigelow, BP, Exxon, Aramco, Bechtel, Fluor, Pharma - Glaxo, Gilead, Rio Tinto Mining, Pilot Co. transportation service centers	Bayer AG (Monsanto), Halliburton, P&G, Kraft Heinz, FedEx Logistics, Mersk, Cosco	Mitsubishi, Foxconn, GE, Marriott, Virgin, Tesla, Toyota	Blank Intentionally

for the new cosmos economy. Each phase lends itself to optimizing a different business strategy with unique strategic goals, cost requirements, and customer profile. Each successive phase builds on the commercial platform and operating infrastructure created by prior efforts made by investors, entrepreneurs, and sovereign state actors.

Connecting the Dots

The cosmos economy is, by definition, a platform for major disruption and innovation. The people and organizations planning to play in this unexploited sandbox get to make up most of the rules and therefore get to pick up the marbles. To develop something new, like a new enterprise in the space-based economy, it takes someone who recognizes there may be setbacks along the way, risks taken, strategic plans revised. It often boils down to courage and commitment.

If this book was a work of fiction the main character would represent human aspiration, the antagonists would be all the obstacles thrown in the way of economic and social adoption, the love interest would be a sense of vision about the future, and the developmental arc of the hero would be the transformation from an eons-old parochial view of humanity to an acceptance that mankind is a space-based species.

The hero (humanity) would have to overcome the pressure of naysayers and opponents who see the potential of space as a threat to their own economic and political power. Each side would struggle to maintain their vision amidst a setting of imminent risk and likely death: a death of spirit, reputation, or even loss of life. And they would acknowledge the importance of commerce as the foundation for establishing new space communities. The hero would define success in transformative terms like the ability to create viable space settlements that would attract like-minded pioneers, the commitment to satisfy investors with significant financial returns on their investments, and the passionate desire to demonstrate the economic, and social, benefits of the space economy. The hero would likely measure their degree of success by evaluating their progress in terms of social and economic stability, whether the space sector has become a well-integrated platform for providing goods and services, and if new emerging trade centers and hubs begin to grow in importance and challenge those on Earth.

At the end of the novel I would have the hero (humanity), like Odysseus, heft a metaphoric oar (e.g., community, commerce, or rule of law) to his shoulder and strike out across the uncharted solar system, away from the newly civilized realm of space settlement, in search of newer frontiers to explore and settle. Stopping only when he's gone far enough into the wilderness of space that no one he encounters recognizes the true nature of the unique burden he carries. He (and we) must go on to the next frontier, and the next frontier after that. From now on there will always be new frontiers to explore and settle in space.

Part VI

Lessons Learned on the Way to the Cosmos Economy

This section summarizes the major themes of the book and lists the major lessons learned from researching the key topics of the book.

30

A New Generation of Pilgrims

There are many visions for leaving Earth to settle tomorrow's space; many reasons to uproot from an old life and pursue a new calling. Some see a life of adventure in space, others see financial opportunity. Some seek a lifeboat from a planet in turmoil with rising seas, doubling population, dwindling ability to feed a growing population, geo-political tensions, greater disparity between rich and poor, and the prospect of never-ending pandemic waves ravishing the globe. Others simply seek the prospect of profit and power. Some will look to space through a pragmatic lens. What reward, they may ask themselves, is there to be gained from the risk of embarking on a perilous space venture. These are the hard-core entrepreneurs and industrialists who envision the new frontiers of space as greenfield opportunities. A virgin frontier ripe for new enterprise. There is a myth about entrepreneurs that they are gamblers, risk takers. This is not so. It turns out that successful entrepreneurs are risk averse. They engage in new ventures only after thoughtful research and deliberation. Their desire is to reduce the risk of failure and increase the prospect of success. The successful businessperson sees both the upside strength of the new venture and the downside pitfalls. It is this same deliberative attention to data and detail that will attract savvy business folk to launching space settlement enterprises.

They will likely be personally motivated by the same mix of intrinsic and extrinsic persuasions that engage their Earth-bound counterparts. Many will be energized by the hope of striking it rich on the space frontier. Some will be motivated by the challenge of being the first to accomplish something that many consider unachievable. Others may be motivated by the seduction of praise and recognition; gaining a reputation as the person who successfully cobbled together a chain of energy generation stations to supply asteroid miners, or who set up a capital equipment leasing operation to supply manufacturers, or who built a brand providing logistics services for others operating in space.

There is, of course, the age-old adrenalin rush that comes from being competitive, of taking on an adversary, defeating unanticipated problems, innovating solutions, and establishing a clear reputation of excellence. Competition is a powerful enticement and is often its own reward. On the flip side of praise is the looming threat of public failure. Some engage because they fear the beast of failure and want to litigate their commercial plan to prove they have something better than the common notion of quality or value. They know they may suffer the death of damaged reputation if they fail which is motivation enough to fight with the fates to establish a commercial success. No guts: no glory!

There are intrinsic motivators, too. Personal values that motivate some to become a better person by acquiring more knowledge about themselves, about business, and about what it means to take a risk and succeed. These people may see the opportunity of space as a chance for spiritual growth as much as a platform for business success. Others may see creating a new venture in space as a path to earning recognition and respect, or to gain a sense of accomplishment where the chances for achievement back on Earth were limited. And like many who came to the American new world there are those who will be motivated by rehabilitation—a chance to redeem some past misstep—for them coming to space may be a way to start over again with a clean slate and to remake a tarnished reputation. Beginning something new in space is an opportunity to build a personal legacy, a way to validate one's worth as more than a nameless stranger in a strange land.

Charting a Course to the New Space Normal

How long it will take, or what efforts will be needed, to become accepted as a full-fledged citizen of a space community is a matter of social acceptance. Immigrants arriving in the U.S. fresh off the boat from "the old country" were easily identifiable by their unmistakable foreign accents and their unfamiliarity with American customs. It was an accepted rite of passage for each new group arriving in the new world to disparage the next new group of refugees and seekers of a new life for their lack of social skills, their tight-fisted pecuniary habits, and their struggle navigating American colloquialisms. Looking down on the new arrival was a way for an earlier immigrant to signal that they had graduated from the bottom of the ladder, that they weren't the naïve *greenhorn* anymore.

So too, it will probably be the same for all the newbies trying to acclimate and fit in to their new adopted home in space. They will face the challenge of figuring out the local jargon, the local etiquette, the local pecking order, and social hierarchy of insiders as they try to make a new life in space. In time, the greenhorn will learn that talking about how things used to be done back on Earth falls on deaf ears—because what matters most will be how things are done now in space. In time, the space pilgrim will learn to navigate the new normal of his or her new home and will give up old affiliation to the life they had before. Their new life, the life they've chosen to live in space, is the only loyalty that will matter.

Being born in space, a true space native, will undoubtedly carry some social clout. Just as being a native of New York or L.A. means that you are more likely to stick around for the long haul and not pick up stakes like lots of out-of-town hopefuls when things get tough or when dreams of instant success don't come true as expected. In fresh-start towns like San Francisco and Los Angeles the locals are hesitant to form full friendships with recent arrivals. Making an emotional investment in a relationship takes a great deal of time and trust that may be wasted when the new arrival gives up and decides to move on to the next town seeking opportunity instead of sticking around for the long haul. The space frontier is likely to attract the loner, the dreamer, and the adventurer seeking some new locale to launch a new identity. A place to build a new brand and then move on to find the next new chance at making his or her dream come true. Regardless of the social dynamics, the fundamental underpinning of space for the next several generations will be trade and commerce. Those who come prepared with commercial skills and experience will have a better chance at success. And those who build a solid plan for the long term are likely to succeed in the new space settlement economy.

Future Careers in Space

For now, most of the jobs in space will be here on Earth. If you are an entrepreneur running your new space start-up, you will run it from here on Earth. If you are an analyst turning raw data collected in space into usable information for customers, then you will do your analysis from here on Earth. Likewise, if you are a VC funding a space venture, if you are a rocket builder, if you are marketing space tourism. Space jobs, with few exceptions, will be here on Earth because most of the doings out in space will be accomplished by robots for decades to come.

Planning a career today, either here on Earth or in space, isn't easy. The structure of our economy and the nature of work are changing and the entry ticket to the party is an ability to keep learning, to adopt new skills, to pivot to new opportunities, and to build relationships with all manner of people. Without a clear picture of what the future of work is going to be, it is impossible to precisely predict the skill set, the core competencies, needed to stay employed. To make this issue more compelling, consider that roughly two-thirds of children entering primary school will find work in jobs not yet created. The pressure to educate the workforce, from childhood through career-changers, so they can compete and stay viable in the labor market if possible, has become a critical societal issue. This is especially true for future leaders.

Space workers will not come cheaply. Critical skills acquired while working for one space company may not be as transferable to another company as on Earth. A highly specialized set of competencies not only presents recruitment challenges but also demands a high salary threshold. If an employee is stationed in space, has buyer's remorse, and decides that they don't like the job or the company it will be very hard to unwind the employment contract. It is likely that contracts will have stipulations that commit someone posted to space for the duration of their tour of duty. The best remedy for this is an extensive vetting process. Also, given that much of corporate space activities relies on proprietary technology, it is probable that an employee will be bound by a non-disclosure agreement (NDA).

A Template for an Innovative Leader

Here is an edited and much simplified job description that represents what is expected of a typical space organization leader. Note that this is not an inventory of the things the leader should do but describes the character and values of the candidate. This is based upon several senior job profiles for well-known international organizations. I have included this job description because it represents what will be required of someone leading tomorrow's competitive organization. As described below, the new space leader must …

> Have passion for serving customers and all constituents with deep commitment to fulfilling and promoting the Organization's mission, the next leader must be a highly engaged, visible, and collaborative leader with a deep understanding of the evolving challenges facing the commercial space business environment on- and off-world. The next leader will show confidence, courage, and conviction in leading the organization and will be energized by the unique possibilities for leadership

and innovation afforded through our firm's outstanding reputation and resources. The next leader will possess many of the following qualities and qualifications:
- An innovation mindset, an entrepreneurial outlook, and the ability to conceive of and implement an exciting vision, in collaboration with others;
- Decisive, objective decision making that moves processes forward in an environment open to change and innovation, as well as sensitivity and ability to productively engage with the company governance processes;
- A collaborative orientation and an openness to building and sustaining synergies and win-win partnerships within a highly competitive industry setting;
- Experience conceiving marketable and timely program initiatives for rapid deployment in a constantly changing market environment;
- Possess a deep interest in engaging with diverse constituencies, including internal and external customers, suppliers, regulators, and changing technology requirements.
- Have a commitment to technology as a tool to bring about innovation and to elevate our level of competitiveness;
- A sophisticated understanding of marketing and an ability to engage in in-depth, strategic dialogue about marketing and branding;
- Strong administrative, management, and leadership skills and an ability to build mutual trust and inspire leadership in others;
- Fiscal and budgetary acumen;
- Cultural sensitivity and a strong personal commitment to diversity and fostering a culture of inclusion and belonging including an ability to plan and manage fully integrated AI and robotic work inputs and outputs.

The successful candidate will possess advanced academic and professional credentials (terminal degree preferred) and a minimum of 15 years of substantial professional experience. Commitment and passion to meet the unique customer needs of our organization is required (Kasriel 2018).

That's a lot! What differentiates this job description is the heavy focus on people skills. The manager of the future is someone who excels at dealing with people at all levels of the organization, not just finding solutions to complex problems. The good news is that most of these "soft" people skills can be taught. Large corporations spend significant portions of their executive training budget on courses like Emotional Intelligence (EQ) which helps empathetically challenged technical folks (quants) to navigate social cues and engage their more human side with others. Executive coaching is now standard fare for mid- and senior leaders and no longer carries the stigma of remediation. Coaching isn't about "fixing" a dysfunctional manager but focuses on how to be a better leader, how to improve the executive skillset, how to engage empathetically with key stakeholders, and how to better manage the career ladder trajectory.

Getting Connected to Space Business and Space Jobs

As investments in new space companies increase there is a growing need for qualified talent to help current and new firms grow and succeed. Space Angels, a leading venture capital firm that specializes in the commercial space sector, launched a website, Space Talent (2020), that connects space companies with the human capital talent they need. The site lists job requisitions for nearly 150 space companies, with positions ranging from engineering to management and marketing. Well-known companies are represented like SpaceX, Rocket Lab, Blue Origin, and Virgin Galactic as well as satellite operators like SES, Maxar's Digital Globe, Planet Labs, Spire Global, and ICEYE.

Women and underrepresented minorities are finding employment opportunities in space start-up companies. This is evidenced by the increasing number of women and minorities who are space organization leaders, senior managers, and board members.

Note that median pay in space jobs is currently a bit less than major IT firms in Silicon Valley. One advantage of the space industry is that space firms are not located in only one concentrated region and are dispersed all over the U.S. in places like the San Francisco Bay Area, Seattle, Denver, Washington DC, and the Los Angeles basin as well as many foreign nations.

References

Kasriel, S. (2018, November 8). *The future of work won't be about college degrees, it will be about job skills*. CNBC. Retrieved from https://www.cnbc.com/2018/10/31/the-future-of-work-wont-be-about-degrees-it-will-be-about-skills.html?__source=twitter%7Ctech

Space Talent. (2020). *Space talent*. Retrieved from https://www.spacetalent.org/

31

Questions and Answers

Lessons from Emerging Markets

By briefly examining the nature of past emerging markets, perhaps we may gain some insights about how the early growth of the future space economy may similarly develop. Are there lessons we can learn by focusing on the enabling and inhibiting factors of less-developed economic centers as they attempted to grow and become more sophisticated markets? Table 31.1 captures an abbreviated inventory of common enablers and inhibitors for Earth's emerging markets and for the early frontier (phase 1) of the space economy. The continuing emergence of new frontier space economic centers may prove to be the vital economic driver of the space economy.

The five phases of industrial space adoption differ from the more familiar Earth-based economic development process primarily because economic development on Earth, historically speaking, has been able to grow and evolve as part of a dynamic competitive system, through lengthy interaction with other viable pre-existing economies. This is not yet the case in space. Space is without a robust infrastructure, there are no entrenched space-based competitors, there is no local labor pool or established technologic resources from which to draw, and there are no ready-made individual or corporate consumers of goods or services. Space is not only a physical vacuum; it is an economic void as well. But not for long.

Table 31.1 This comparison demonstrates similarities of emerging markets on Earth and the Phase-1 space frontier. There are also distinct dissimilarities such as rents, labor costs, the impact on indigenous peoples (on Earth), and the scope of competition especially at initial stages of development

Common factors of earth's emerging markets and undeveloped space frontiers	
Enablers	Inhibitors
Ability to leapfrog old technologies with robotics, innovation, and artificial intelligence	Limited local customer base for consumer or industrial output—especially at first
Opportunities for start-ups. Incentives to locate	Transport and shipping challenges
Access to raw materials and production resources	Limited skilled labor and/or production resources
Limited competition enables premium pricing	Limited access to business services resources
First mover advantage attracts investors/partners	Unattractive quality of life
Limited initial local/regional competition	Undeveloped or marginal infrastructure
Viewed as potential "acquisition incubator" by larger, better capitalized tier-1 and tier-2 firms	Challenge of navigating complex and often contradictory regulatory landscape
Early market entry presents learning curve advantage benefits	Local supply chain may be inadequate for business needs thus increasing costs
Investment is typified by traditional industries like mining, manufacturing, agriculture, and general trade	Lack of availability and reliability of infrastructure and/or utilities may cause operational inconsistencies and inhibit production efficiencies

Lessons from 21 Questions and Answers About the Space Economy

Here is a list of some of the main research questions I explored while writing this book. When I started to research and write this project, I didn't expect to find definitive answers to many of these questions, but it turns out I discovered much more than I originally hoped. I inserted these questions in the introduction near the beginning of this book to encourage reader interest and curiosity. If you ventured to answer these questions, then take another look to see if you've changed your thinking or if you learned to think differently about some of these topics.

Here is a summary list of most frequently asked questions, along with my responses, from presentations I've given on this topic and from the give-and-take discussions during the interviews I conducted for my research.

1. When will the new space economy officially begin? 2040? 2100? Now? Never?

- The new space economy has already begun. It is expected to kick into high gear by mid-century or sooner.

2. What is the prevalent/dominant reason for humans to go to space? What is the "right" reason? Are these two perspectives in conflict or in agreement?

- The visions about space range from escaping a hopelessly ailing planet to moving all the environmentally threatening industrial activities off planet to let Earth heal and rejuvenate. Some see space settlement as humankind's noble manifest destiny. No matter the rationale, each vision has some kernel of merit, a common thread, that posits going to space is for the good of the planet and the betterment of the species, an altruistic desire to repair and heal.
- Underlying this benevolent theme is the pragmatic desire for profit and power in the new space economy.

3. What is the downside—the reason(s) to reject—the coming space economy?

- The new space economy will naturally be a potential threat to established industries and nations on Earth who will resist space economic (and political) development to protect their current standing. Nations who are not spacefaring will see space as the new military threat and will either resist space activities as an act of defense or align themselves with a federation of space nations and companies. There will be national rivalries for control of resources and optimal locations either on planet surfaces or in deep space. Self-interest will determine the level of support or resistance.
- A major operational inhibitor is the corporate short-term planning mindset. Space is a long-term endeavor. Companies will have to give up the notion of short-term quarterly reporting as way of determining success and signaling investor guidance.

4. What role will humans have in space?

- Humans will have a minimal role at first because robots and other tech will be used to construct outpost sites and conduct industrial operations. Humans will act in a supervisory role via teleoperations.
- After initial infrastructure is constructed humans will begin to establish space settlements and outposts in greater numbers. By approximately 2100 there will be a thriving human population in space.

5. Are prevailing notions of people living in space settlements realistic?

   - In time, as settlements are established, humans will have a greater role.

6. Given the hostile nature of space will robots or humans do most of the work? There are two prevailing views of how humans will participate in space.

   (a) Humans will play a dominant role long term. But near-term human activity in space will be limited because of our vulnerability to radiation, the effect of reduced gravity, and other life-threatening impediments. Controversial research is exploring how current human biology can be modified to best fit with the space environment and turn Earthmen into Spacemen.
   (b) The prevailing view is robots will continue to do the industrial heavy lifting as well as most of the manufacturing, mining, and other operational tasks. People will continue to manage robotic operations and define success. Robots and people will work collegially as cobots each providing complimentary skills.

7. Will there be a time when dominant economic activity will shift from Earth to space-based markets?

   - It may not be so much an issue of economic dominance but, rather, an issue of economic diffusion throughout key commercial hubs in the solar system. As the new space economy matures, and intra-space commerce increases, Earth's dominant role in the greater cosmos economy will diminish.

8. What will be the impact of the vast abundance of space resources on Earth's global economy?

   - Precious metals will no longer be as precious. Critical raw materials (from minerals to water) will become easily available commodities. Gold may become the new aluminum.

9. Other than mining, manufacturing, tourism, energy generation, and ancillary supporting operations, what will be the dominant industries in the new space economy?

   - Different industries will flourish at different phases of economic adoption.

- By the end of the current twenty-first century finance, investment, energy, general trade, logistics, agriculture, health care and pharma, heavy and light manufacturing, and a collection of industry segments not yet created will provide the greatest return for investors.
- Initial business-to-business (B2B) commerce will eventually be eclipsed by increasing consumer-based transactions as humans migrate to space in greater numbers.

10. Will the economic development of space conform to a universal master plan or will space develop organically, driven by the priorities of countries and companies in pursuit of their own vested self-interest?

    - A master plan is highly unlikely. Self-interest and economic market forces will determine how national or corporate space activities will take place. These numerous plans and visions will likely conflict with one another.

11. Is there a predictable series of adoption phases that may be used as a guide for a company planning the best time to enter the new space economy?

    - The five-phase adoption model patterned after Everett Rogers (1962) will help investors and others better understand when to enter the new space economy with investments or new ventures.

12. Are there business models best suited to each adoption phase that will enable a company to better structure their business plan?

    - Not all business models will work effectively at each phase of adoption. It will be critical that a company consider not only the timing of market entry but also the type of business model in support of the business plan.

13. Where will we create new communities in space? Will the greater space economy expand across multiple locations in the solar system?

    - Once the initial frontier adoption phase demonstrates the viability of the new space economy (as a live case proof of concept), the industrialization of space will rapidly spread throughout the solar system. Controlling this growth will be impossible.
    - Expansion of free space cislunar communities in the goldilocks zone of the Earth's orbit is a likely early phase. Some communities will be robust

commercial centers and others will be industrial complexes. These space cities will become gateways to different areas of the solar system.

14. How will the development of the space economy define the relationships of sovereign nations with each other and with dominant space enterprises? Will future space partnerships ignore or reinforce the implications of traditional political borders?

- The prospect of recalibrating the balance of power will enable nations and private enterprise to form alliances that will redefine old traditional political boundaries and confederations. China will likely dominate in space in the coming years, causing political tensions for decades to come.

15. Will supply chains develop independently (organically) or will they be choreographed by organizations wishing to control the direction of the new space economy?

- Yes and No. Corporate controlled vertical integration will support customized supply chains. But more competitive independent suppliers will establish open channels in keeping with current and evolving levels of demand.

16. Will space-based technologies impact (disrupt) longstanding Earth-based business?

- New technologies and processes designed for space will quickly migrate to Earth's commercial business sector. Because adoption of new technologies will be critical to stay competitive traditional manufacturing will undergo a radical overhaul to compete in the new transglobal economy.
- Reverse migration of biotech developed to enable long distance space travelers by altering their genome or augmenting body components will create a new medical consumer market of medical practitioners peddling a complete menu of replacement and upgraded body parts that will help to enhance health, reduce the effects of disease, and prolong life. The impact on the current traditional health care industry will be enormous.

17. How will treaties, accords, and other well-meaning agreements be enforced in space? Who will enforce and adjudicate disputes in space when the space laws of one country conflict with those of another country or with the corporate rules/policies of a private enterprise?

- Enforcement of space law is problematic. Given the combination of overlapping and contradictory corporate policies, national space laws, and great distances it may be a long time before there is a universal approach to space frontier justice.
- You didn't really think human nature would be any better in space, did you?

18. Will the new space economy eventually become independent of Earth's economic support? If economic independence is reached, then when will this occur? Will Earth and space become economic competitors?

    - The multiple sectors of the new space economies will undoubtedly become independent from Earth. A loosely "federated" solar system economy (or economies) will be driven by the emergence of independent economic centers and communities that will lead to a decentralized new space economy.

19. Will it be today's engineers or tomorrow's managers who will lead the way to a fully formed space economy?

    - This is not an either/or answer. Both engineers *and* managers are critically needed to launch and maintain the new space economy. In time managers will determine the economic growth of the new space economies. Space is big enough for both quants and poets.

20. While much science fiction literature assumes a well-integrated and robust space-based economy, is there something to learn from sci-fi about how the space economy will develop?

    - Science fiction has a general tendency to see future space economics in terms of post-capitalism. A recurring dystopian narrative supports the idea of a few mega-corporations in control of space where individuals would be identified by their corporate citizenship instead of national affiliation.
    - Most science fiction subordinates women as well as underrepresented minorities. Companies in the new cosmos economy are committed to rejecting these past prejudicial practices in favor of rewarding talent and professionalism based upon merit instead of superficial stigma.

21. How can individuals best plan for a career in the space industry?

- Opportunities for both engineering and managerial professionals are dramatically increasing in the new space economy. Learning how to effectively manage people, projects, process, budgets, and/or brands will be vital general skills in the new space economy. The most critical skills are relationship management and the ability to make decisions using both rational and intuitive inputs.
- Space is an industry with high demand for qualified women and other underrepresented professionals. There are many ground-floor opportunities for a long-term career in the new space economy.

32

Lessons Learned

The new space economy is likely to happen much sooner than expected. The critical mass of technology, economic demand, and investor interest are pushing the industrial space growth curve steeper and steeper. It is only a matter of time before settlements and outposts—either funded by countries, companies, or private investment cartels—take shape and begin to establish a foothold for man's permanent presence in space. There are some speed bumps that are known and probably many more that are yet to be recognized, but the momentum of investment and innovation is quickening. What remains is a magic list of the things we don't know we don't know. That will come in time.

This inquiry started with a simple nagging question:

What would people living in space do all day?

When I started this project I assumed that all I had to do was talk with people who were in the space industry to learn about their long-term plans and to research articles that were written about what the future would look like when humans settled space. I just assumed there was a lot of hard thinking about this topic. It turns out that most of the hard thinking has been slanted to the technical requirements of going to space (very important) with lots of speculation about how to build habitats on Mars or self-sustaining cislunar stations to service travelers between Earth and outworld destinations. It also turns out that not much was written about the roadmap to a space economy; who would traverse that roadmap, or what value that new economy would bring to the current global financial structure in order to justify the cost as a return on investment.

I had heard that people were going to space to live and work, but I could never pin down why they were needed in space—what their added value was—or the type of work they would do out there. The major lesson I learned from asking this question was that no one else had the entire answer. People could see centuries into the future where there would be countless colonies, outposts, and settlements in space but the middle ground—what happens next, what happens tomorrow—was a magical black box of conflicting visions about the what, why, and how of space settlement.

I learned that many of the firms doing business in space today will not likely persist into the future unless they pivot with agility and reorganize their operations to match a constantly changing cosmos economy. A traditional business plan based upon large-scale consumer subscription for internet services may work well on Earth where there are lots of potential consumers but, for the foreseeable future, there are no consumer in space to support that business model. Likewise, planning to outfit every space family with a stove, refrigerator, washer, and dryer just like on Earth would fail to recognize the probable Kibbutz-like structure of early space settlements where most domestic activities like cooking, dining, and cleaning will be done communally, thus eliminating the need for individuals to own all those durable goods. I learned that for the space economy to grow and mature it must become more than one big strip mine. Refining ore pays off best when there is an integrated industrial base to put all that raw material to commercial use. Just because there is an apparent endless supply of valuable raw material in asteroids doesn't mean there is endless demand for manufacturers to produce an endless stream of goods for customers and consumers who will have to purchase all that production. What good is the prospect of endless supply when there is no assurance of complementary demand?

I learned that as more gold is mined, the less valuable it becomes. As the supply of precious materials goes up, scarcity goes down, along with the value and price of the commodity. The abundance of raw materials, minerals and water in the solar system is a wonderful thing – unless you are a commodities speculator.

I learned that some people think that building a space-based society will serve as a lifeboat for our troubled planet Earth. Others are more optimistic and see space as humanity's manifest destiny: the obvious next step in our species' evolution. Along the way, I stopped taking for granted that everyone I spoke with shared a common vision about the economic potential of space. In fact, the lingua franca of space had more to do with Star Trek and Star Wars than with Dow Jones and NASDAQ.

I learned that Public-Private Partnerships may transcend our current parochial notion of political boundaries. That is, space may enable strange bedfellows where a private firm based in one nation may partner with a different national government. Might SpaceX become a partner with South Africa? Could Blue Origin partner with Japan's JAXA? Might a Chinese manufacturing company like Foxconn partner with the U.S. government? Strange bedfellows.

I learned that the human animal is supremely ill suited for a life in space. While some may worry about man's impact on the pristine environment of space, a greater concern is how space will impact human biology. More than deadly gamma rays, the lack of readily available breathable air, the need to BYO food and other vital necessities, or the emotional impact of a lonely life of confinement, there has yet to be solid evidence that our species can successfully reproduce off planet. Capt. Henry Schwartz, M.D. (Ret.), former physician to the U.S. Navy Pacific submarine fleet, noted that the psychological effect of long-term isolation takes a temporary toll. The promise of returning to port and normality, always present, is a strong balancing palliative. But space pilgrims who sign up for a one-way trip to a far-away space outpost won't have the restorative luxury of imagining a future back on Earth with family and friends. How will they cope with a life sentence of separation?

I learned that there is an extremely limited capacity for shipping goods from space to Earth. This constraint, in effect, is a fork in the road to space building a space economy. If the revenue and capacity of shipments from space to Earth are held to a trickle, then the design and consumption of goods produced in space will be directed to space-based customers. But if the ability to ship very large tonnage of goods from space to Earth (generating healthy revenue) is resolved, then Earthly customers will steer the course of the space economy.

I learned that for all the apparent opacity of a common vision, there is a sincere expectation of high returns on investment soon for the new space economy. To buttress the investment expectations inherent in these studies, today's definition of commercial space (i.e., satellites, communications, data analytics, etc.) has started to evolve from activities exclusively below LEO (low Earth orbit) to ventures on the Moon, asteroids, Mars, and other locations in the solar system far beyond convenient Earth markets.

I learned that the potential impact of an industrialized solar system may disrupt our current notion about our home planet's future dominance in civilized space. Space has virtually limitless resources, unlimited room for industrial production, and plenty of room for new space settlements. Space also provides the promise of unfettered independence for those individuals and communities who seek to travel beyond Earth's traditional political orbit. At some future point our Earth home may be just another tourist destination offering a taste of the old country with an opportunity to experience higher gravity. A theme park vacation destination for the family.

I learned there is some lack of agreement about who will maintain law and order in space. Treaties, accords, and agreements aside, corporate interests and national desires for political control and financial gain will likely compete for dominance with each other just as on Earth. Space will be at once an open range and a setting for contested claims of ownership and control and of influence and military supremacy. When the cavalry finally comes to the rescue, the design of their uniforms may not be what you expected.

I learned there are passionate feelings about *human* space settlement. (An editor of a space organization magazine told me that his readers would be in revolt if he carried a story about robots replacing humans in space.) This ardent belief that people will populate space is widespread and is an underpinning tenet of today's space culture; it is heretical to suggest that humans may likely play a subordinate role. One conclusion I've drawn is that, for some space zealots, the fantasy of living in space—likely fueled by heavy doses of science fiction—is stronger than the actual desire to go there and scratch out a new civilization.

I learned that automation and AI will dominate the first phases of space development. Space is truly no country for Earthmen and robots will clear the way for humans by constructing outposts and bases on the Moon or in orbit. Space is not a place for traditional human manual labor and is not a place for the lessons in management and leadership taught by business schools today. I learned that when you remove people from the equation, then production in space becomes more efficient and even more cost effective. But creativity and innovation suffer.

I learned that the tsunami of space-based commerce will crash on the shore of our global economy with disruptive force. This will change the nature of Earth-based business as much as it will establish space-based firms as leaders in their economic sectors, as innovators leading the way in the development of the next industrial stage. I learned that the drivers and variables are constantly in flux; that no one has a clear crystal-ball vision of what space will be because no one can accurately foresee the technologies at play nor the full range of potential benefits that will come from creating tomorrow's space settlement economy. We just don't know what we don't know.

I learned space is much more than well-known companies like SpaceX and Blue Origin, much more than asteroid mining outposts and settlements on Mars. It is new university programs offered in space commerce, space ports popping up all over our planet, and billions of dollars of investment pouring into the commercial space sector. I learned that the new space economy is a wide-open frontier ready-made for new ventures and entrepreneurs as funding shifts from government programs to private capital.

I learned there are plenty of career opportunities for engineers, program managers, administrators, and educators. I learned that the space economy isn't a fringe business idea anymore, it's a robust and dynamic business sector with plenty of room for new ventures willing to stake a claim and make an impact on the future of human space settlement. I learned that the people working in the space sector are creative, dedicated, and welcoming to new ideas and new people. I also learned there is an open invitation for like-minded persons to join the ranks of the many new space companies dedicated to bringing about the cosmos economy.

Finally, I learned that the new space economy isn't confined to the future, as I originally thought when I started to research this project. The space economy is here on Earth showing up as U.S. government projections of a multi-*trillion* dollar economy by mid-century, it is the billions of dollars of annual investment capital, and the hundreds of new start-up companies competing for space contracts and customers all over the planet. It is now.

There was a Time

There was a time when voyaging to the New World was the province of shipwrights and sea captains, navigators, and weathered sailors. There was a time when navigating the uncharted expanse of the Great Plains was the domain of wagon masters and experienced prairie scouts. There was a time when charting the course of the transcontinental railways was the domain of surveyors and day laborers. There was a time when immigrants and the disenfranchised seeking an opportunity could find their place on the new frontier.

But the sea captains and wagon masters were guides, not settlers. The shipwrights and surveyors had little interest in making a new life in the wilderness. They were all enablers of the future, the engineers and technicians who cleared the way for the immigrants and visionary pioneers. The merchants, ranchers, and farmers—the pioneers and seekers of a new life—were the settlers of the New World.

The builders of ships and wagons made the great migration happen because of their technical skill. Once the ocean had been navigated, the wagon trails carved out, and the pioneers transformed into settlers the role of the technicians changed; eclipsed by the vision-makers and the entrepreneurs who followed. Frontier settlement is the story of settlers who took the big risk to step into an unknowable life. Such is likely to be the story of space.

Soon living in space will be commonplace. The technologies of air travel, automobiles, internet, mass communications, and industrial agriculture were once new and disruptive but are ordinary now. Human settlements in space will be just as accepted and routine.

Today we are fascinated with stories of rocket makers and the high-tech breakthroughs they create. Tomorrow we will warm to the personal stories about pioneers and settlers who made the trek, took the risk, crossed the cosmic seas, and built a new life out on the space frontier.

33

How to Be Part of the New Cosmos Economy

The real story about leaving Earth and venturing into space isn't about the tech stuff. It's not even about courageous adventures or boldly going to brave new worlds. It's about the profit stuff. It's about private investors, big business, and entrepreneurs who are creating our civilization's next industrial revolution in space. Naturally, there is uncertainty and risk along the way. But uncertainty goes together with two underlying truths of business: (1) High risk enables high reward and (2) risk avoidance guarantees there will be no reward. Entrepreneurs will seek to create new businesses in space because the prospect of high stakes rewards for the risk takers is just too good to pass up. The inevitable failures are just collateral damage on the path to defining the new transplanetary economy. Sometimes it all boils down to a matter of courage.

Why Is the Cosmos Economy Different from all Other Economies?

Here are some hints. The new space economy is …

(a) A completely blank canvass for adventurous entrepreneurs upon which to build their new enterprises.
(b) A place where business rules are yet to be written or enforced, if ever.
(c) A place for creative business visionaries to innovate and define success on their own terms.
(d) An economic platform of endless frontiers loaded with almost limitless resources.

(e) An engineer's playground of endless technical possibilities.
(f) A VC investor's dream come true!

The prevailing investment argument about the cosmos economy holds that in order to reenergize the world economy we cannot overlook the greenfield opportunities that come from our expansion to other worlds. This is not about advancing science although that is part of the process. Neither is this about the military strategy of claiming the high ground of space although there is no denying that is also baked into the formula. This is not just about the fascinating high-tech developments turned out by leading R&D labs around the planet although building a new space-based economy will rely on superior engineering ingenuity. This is about the business of space-based capitalism. Either we successfully move our planet's civilization into the new territories of space or by failing to make the leap we sink back into the morass of our present global parochialism.

Fasten your seat belt, it's going to be a fantastic ride.

Now It's your Turn

The original motivation for this project was to chronicle and explore the activities going on at the very beginning of the new cosmos economy: the budding business ideas, the innovative thinkers, the motivating factors, the risks taken, the failed attempts, and the stellar successes. I soon realized the complex space economy I saw was only a glimpse of the future. My dad used to say that knowing the future was like capturing lightening in a bottle, an impossible task. Grasping the full essence of the new space economy is just as illusive. There is no way to know of all the corporate plans or the twists and turns that occur at the beginning of something new and powerful. I've tried to add some structure and form by suggesting the five probable phases of economic adoption and matching these progressions with likely business models, but I recognize that this is only a best guess. Because you just don't know what you don't know.

Extra Credit …

Maybe it's my years as a college professor but I have a homework assignment for you. I'm asking you to be a contributing historian of the new cosmos economy. If you are an active investor, a space business entrepreneur, a board member of a company in the space business supply chain, an engineer developing something new that will get us all closer to establishing our civilization in space, or more importantly, if you are an average adventurous observer who is passionate about space, my plea is to keep notes on developing space events. This is an opportunity for you to be part of human history. Adding your personal voice to the developments of the next few years may have a significant historical impact on how future generations see the contextual seeds of their future civilization.

But you may say that you don't work in the new space industry. Maybe this is your time to change careers to keep that dream alive. There's plenty of room for people with a passion about space. Make the leap.

The noted historian **Becky Erbelding,** when chronicling the rise of the Third Reich and the subsequent Holocaust in her book *Rescue Board* (2018), noted that much of what we know today is the result of historical research from personal journals written by average individuals who had a ringside seat to history. More than just capturing raw information she also envisions the future presentation of an individual's journal materials in a museum display and suggests four very simple steps. The following steps are quoted from her method:

1. "Take 10 min at night (or however long) to free form write. With a pen if you have good (non-cursive) handwriting, otherwise type."
2. "If you type, print it out on stationery. SINGLE SIDED ONLY. Exhibit designers can't use double-sided pages."
3. "Write what happened that day, then where you were when you saw what happened, what your reactions were, what you felt, what you hope will happen next."
4. "Then save it somewhere safe. That's it. Super simple. You'll be very glad you did it."

Being an historian of the future doesn't demand that you have a PhD in history or even acute predictive skills. Capturing the details of today with the understanding that some of the events may become important in the future is not about being able to draw profound dotted lines between data points that don't yet exist. It simply requires you to have a passion for your subject.

As Soraya de Chadarevian relates in her 2016 paper, *The future historian: Reflections on the archives of contemporary sciences,* being an historian of the future is more about being an archivist than about being a seer. To make this point she begins her paper with the following anecdote.

Some sources make a big splash. A recent example is the letter penned by Crick to his twelve-year old son Michael in May 1953 that described the structure of the DNA double helix before it appeared in print. The letter had been in private hands for 60 years, when his son decided to sell it. It fetched an unprecedented six million dollars, paid by an anonymous buyer. This was more than all the other Crick papers together that were acquired by the Wellcome Trust a decade earlier for what then seemed a very hefty price. The sale of the letter was widely reported in the media and although the original document is held in an undisclosed location, digital copies now pop up on countless internet sites. More usually, historical documents or sources especially of the paper kind lead a more discrete life.

As I worked on this project, I fully understood that the facts and guesses presented in this project may or may not prove valid 50 or a 100 years from now. But the project's purpose will be sustained if you kindly consider my request and start logging the developments of the emerging cosmos economy with an eye to benefitting the work of future space historians. Who knows, your observation or an underlined notation in your daily planner may trigger the research work of some future study about the beginnings of Earth's next major historic leap. Your notes about the development of the new space economy may become the foundation for the next generation of historians and space entrepreneurs.

This is how you can participate in the new cosmos economy.

References

de Chadarevian, S. (2016). The future historian: Reflections on the archives of contemporary sciences. *Studies in History and Philosophy of Science Part C: Studies in History and Philosophy of Biological and Biomedical Sciences, 55*, 54–60. https://doi.org/10.1016/j.shpsc.2015.08.004.

Erbelding, R. (2018). *Rescue board: The untold story of America's efforts to save the Jews of Europe*. Doubleday.

Appendix 1: Research Findings

My research polled a mix of space business professionals, academics associated with the space industry, and people not directly connected to commercial space firms to inject a broader point of view. The purpose of the research, conducted in 2019, was to get a baseline of general perceptions about the business of space in the not-so-distant future. Some of the responses were expected and others were surprising.

The following selection of survey responses helped to shape many of the assumptions presented in this book and helped to guide my subsequent interview questions with space industry professionals and others who were likely to see the potential growth of the space economy. The ages of respondents ranged from undergraduates in their early 20s to senior executives. Although administered via the web, most respondents were in Southern California which has a large aerospace community and thus an assumed affinity for space-related business activities. All survey responses were anonymous.

Of course, there is no guarantee that any of these research findings will hold up in the next decades, but given the rapid acceleration of events coalescing in the commercial space industry sector it is likely that some of the assumptions represented here will maintain validity into the next generations of space settlement and commerce.

When Will the Space Economy Likely Achieve Economic Self-Sufficiency?

Just over two-fifths (42.1%) of respondents envisioned by 2050, nearly one-third (31.5%) pegged this milestone at 2075, and the remaining 26.3% answered that this would take place by or after the turn of the next century (2100).

Summary Comments: Recall that data for this survey was collected in 2019. Many respondents to this question would likely consider 2050 as something she or he might experience in their lifetime. A younger, millennial, respondent would even think of 2075 as a date that would occur within their life experience. Yet over 73% thought that a space-based economy would reach a viable level of economic self-sufficiency by 2075, a very optimistic view.

This closely comports with the widespread belief that mid-century (i.e., 2040–2050) will be an inflection point for space settlement and commercial investment.

Will the Space-Based Economy Rely upon a Mostly Human Workforce in Space?

This question was asked to get a more solid read on the topic of human vs. automated space activities. I had encountered ardent resistance to the notion that robots would pave the way to space instead of humans and I wanted to see if this was a pervasive assumption. Over one-third (36.84%) responded that it was highly unlikely that humans would make up the space workforce and 42.11% indicated it was unlikely. The combined total of 78.95% indicated a strong assumption that robots, not humans, would dominate in space.

Summary comments: I had quickly learned that the subject of humans vs. robots was, to some degree, a lightning rod for debate. The argument for a mostly robotic workforce is reduced to (1) effectiveness (an automated workforce can keep working all 168 hours per week), (2) cost efficiency (unlike people, robots don't require expensive life support, life insurance, food, housing, or other amenities unique to sentient beings), (3) expendability (if a robot breaks down it can be repaired or replaced, but if a human dies or is injured it is a very big deal that may cause a safety review), and (4) mission focus (putting a human in a position of risk shifts the focus away from the mission to sustaining human life which, although understandably necessary, detracts from the foundational purpose of the mission).

When Do you See an Opportunity for your Industry to Participate in the Space Economy?

Surprisingly, nearly half of respondents (46.37%) indicated it was likely that their firm would participate in the space economy as early as 2030. Nearly two-fifths (37.84%) opted for likely industry participation in the space economy by 2050 or later. Nearly one-sixths (15.79%) responded that their industry was likely to never participate in the space economy.

Summary comments: The response to this question supports the assumption that there is a general sense of urgency about potential participation in the space economy. The anticipated 2030 target date is less than a decade away. With nearly half of respondents choosing this date for engagement it may be concluded that there is a general acceptance in today's corporate world about actively participating in the future space economy.

When Will your Children or Grandchildren Likely Live/Work in the Space Economy?

The response to this question was closely split with less than three-fifths (57.8%) responding either likely or highly likely that their descendants may have an active role in the space economy. The remainder (42.2%) clearly rejected the notion their children or grandchildren were likely to love or work in space.

Summary comments: Given the range in respondent ages (undergraduates in their early 20s to senior executives), I was eager to ask a question that related more to personal perception and less about business. This is an area for further research: Why do most respondents believe there will be an active space economy in the 25+ years but also feel it is unlikely that their offspring will likely participate?

How Could Potential Investors Be Incentivized to Participate in the Cosmos Economy?

This question presented a list of ten incentives that might act to motivate a business to start and conduct business in space. The sources for the list options came from research about how governments attract and reward private industry to conduct business in a new domain as well as how Public-Private

Partnerships (PPPs) use incentives to enhance the prospect of entering into a relationship to create a product/service for the public good. (Note: respondents could choose up to three reasons, therefore percentages don't equal 100%.)

The top reasons cited were (1) Early entrants are best positioned to create long-term ROI in space (65.71%), (2) AI and robotics, instead of human labor, will reduce production costs in space (58.82%), (3) Unbridled access to vast untapped mineral and other resources in space (52.94%), and (4) Early entrants get to make up the rules of the game and extract premium prices (47.06%).

The least attractive reasons cited were (1) Impact of climate change will reduce Earth's food production capacity thus creating off-planet opportunities (35.29%), (2) Regulations are difficult (or impossible) to enforce in space (23.53%), (3) Markets now controlled by companies or countries on Earth will likely become competitive with space-based providers (11.76%), and (4) Goods, services, and resources now controlled by companies or countries on Earth will compete with space-based providers (5.88%).

Summary comments: The top reasons aligned well with expectations about the advantages of being an early entrant in a new market, the leverage of employing automation vs. human labor in space, the lure of available raw resources in space, and the how first movers get to define the playing field.

An unexpected finding of this question is found in the least attractive reasons for initiating new business in space. The weight of the survey responses doubted that the new space-based economy would compete with the traditional Earth-based economy. This either points to a lack of confidence in the viability of a space economy or to the assumption that space producers will primarily serve space consumers and customers. As mentioned in an earlier chapter, this is a highly probable outcome.

Why Might Firms, Investors, or Entrepreneurs Be Reluctant to Participate in the Space Economy?

Survey respondents had a clear notion of why *not* to venture to space. The top reasons they felt that space was a bad investment opportunity were (1) Unknowns outweigh traditional business assumptions (58.82%), (2—tie) Hostile environment reduces safety and increases insurance and other expenses (52.94%), (2—tie) Transport and shipping challenges (52.94%), and (2—tie) Undeveloped or marginal infrastructure (52.04%).

At the bottom of this list was (1) Unattractive quality of life (41.18%) and (2) Limited access to business resources (11.76%).

Summary comments: These responses successfully capture the general hesitancy about doing business in an unestablished market such as space. While these assumptions about the risks of launching a new business are quite valid in the short run it is worth noting that a longer-term perspective about adopting later stages of development would mitigate much of the uncertainty embedded in these responses.

What Are Likely Government Incentives for Companies Launching a Space-Based Enterprise?

Considering that Public–Private Partnership business models (PPPs) would play a significant underwriting role in the early stages of the space economy, this question attempted to better understand which government incentives may best resonate with entrepreneurs seeking financial support in the new space economy.

It turns out there was insignificant response to adequately tease out major incentives that might be employed to jump-start new business ventures in space. At the very bottom of the list were four potential incentives that surprisingly garnered no support at all; they were (1) a sanctioned ability to sub-lease concession assets, (2) concessions for dual use of capital assets (e.g., launch and terminal facilities), (3) expansion of IP protections (e.g., licensing concessions), and (4) an inexpensive source of investment capital (e.g., for infrastructure).

On the other hand, the top likely incentives provided by government were (1) R&D subsidies (26.32%), (2) Concessions for exclusive mineral rights in a specific territory (21.05%), (3—tie) Granting a virtual monopoly in a specific sector for fixed term (15.79%), (3—tie) Exclusive or preferential contracts with provider(s) of incentives (e.g., governments) (15.79%), and (4) Concessions on taxes, tariffs, etc. (10.53%).

Summary comments: One of the more attractive advantages of a PPP is the ability of the private sector partner to extract special concessions that would otherwise not be available or too costly to acquire. These concessions are meant to provide the private partner with an eminent competitive position not available to all other providers in the market. The concessions also have the effect of acting like "golden handcuffs" keeping the private partner obligated and connected to the partnership.

What Will Be the Impact of Space-Based Business Activities on Global Commerce?

This question was asked to see if there was a perceived difference between a future space economy that was either fully integrated with Earth's global economy or an isolated space economy that may develop with little dependence on Earth. The response indicated there was a perceived difference between the two options with a bias toward benefit to the current model of our global economy.

Almost three-fifths (57.89%) indicated that the future space economy would be beneficial to the global economy. Nearly one-third (31.58%) perceived the space economy as an alternate/separate economic system from Earth's global economy. Only one-tenth (10.53%) believed that the space economy would have no impact on the global economy. No one believed that the development of the space economy would be detrimental to the stability of Earth's global economy.

Summary comments: It is fair to say these perceptions about how the new space economy will or won't collide with Earth's global economy are currently true. Frankly, there is still no evidence to the contrary indicating the industrialization of the solar system will pose an economic threat to business-as-usual on Earth. Given the economic forecasts about the potential growth of the space settlement economy that range from $2.7 trillion to $3 trillion in the next few decades, and the increases in corporate and national investments in this domain, it seems highly plausible that responses to this question will likely change if asked in the 2030s or 2040s.

How Will Space-Based business Activities Impact Established Commercial Producers?

Nearly seven-tenths (68.42%) saw the impact of commercial space impacting companies most of all. This response is understandable given the general perception that commerce is the exclusive domain of private companies. However, there is no rule that says nations can't operate private businesses. In fact, China's SOE (State Owned Enterprise) model extracts revenue and influence all over the world in a wide range of business sectors.

Summary comments: This question revealed a general lack of awareness about the role of sovereign nations developing commercial capabilities in space. Countries, not companies, have the financial deep pockets and persistence to establish beachheads in space (on the Moon, Mars, cislunar space, etc.) and maintain permanent manned operations. Nations also bring the added clout of military capabilities (e.g., The U.S. Space Force) to ensure these outposts have secure territorial claims. Companies have current limits on these enforcement abilities, although this may change in time, unless they form alliances with other firms and/or nations.

The balance of trade and influence between Earth's traditional global economy and the new space-based economy is a topic that invites further study over time.

What Will it Take for Space-Based Enterprises to Be Economically Viable?

Over two-thirds of respondents (68.42%) defined the future economic sustainability of space-based enterprise to come from a mix of individual consumers and corporate customers. Less than one-sixth (15.79%) assumed that commerce in space would be propped up by government subsidies or other concessions.

Summary comments: Given that NASA (and other national space agencies) have relied upon government funding and support for more than half a century, it is natural to assume this is the normal way to sustain space activities. But this is not supportable in the long run. Space enterprise must be able to persist on its own merits by producing goods or services that provide solutions and meet customer demand. That's the whole point of opening the new frontier of the space settlement economy.

Appendix 2: Recommendations for Future Research

I approached this topic with a less-than-complete understanding of the terrain, learning as I went. Each new conversation prompted new questions that begged answers that would only come with time. As of this writing, much of the thinking about this topic is inconclusive or biased in favor of an agenda, business plan, or sacredly held assumption about human habitation in space.

The following catalog of looming questions serves as a starting-off point for further inquiry about the growth and development of space settlement and commerce. Let me know what you find out!

(a) **Supply chain** dynamics in space. An overview of how future supply chains will develop for different industry sectors, how they will be managed, and how they may interact with and impact similar industry supply chains on Earth.

(b) **Financial transactions.** There have been some comments (Krugman 2010) about the impact of space travel vis a vis time shortening/lengthening as applied to interest accrual, and the potential for manipulative arbitrage, etc. Beyond that, how will financial systems be managed in (1) a frontier space economy, (2) a later highly networked and interactive space economy, (3) what currency species will prevail and how will it be guaranteed, and (4) what will be the [new] standards for credit worthiness?

(c) **Skills preparation.** I was unable to get a clear inventory of the unique [human] skills needed to succeed in the new space settlement economy. I am curious to see if there will be fundamental differences in core capabilities for applications in space compared to current best practices on

Earth. Specifically, in such areas as (1) leadership development, (2) entrepreneurial applications, (3) robot and Cobot working environments, (4) remote management operations, and (5) a balanced blend of managerial and technical competencies.

(d) **Efficacy of business models** adoption to a space setting. Space will present a broad set of unknown variables that may enable or discourage commercial activities. It will fall to a future researcher to examine how today's business models hold up over time and to document new forms of business and the context that supports their success or failure.

(e) **Sociology of space outposts and settlements**. As space communities and enterprises progress through the phases of adoption it would be worthwhile for a future researcher to chronicle the social aspects of communities in space. How will corporate sponsored communities differ from independent space shtetls and those that were politically established? A corollary dimension will be how these different space communities interact, settle differences, and form alliance for commercial or military purposes.

(f) **Colonialism**. Will sovereign-sponsored settlements express as a new colonialism? If so, how will this colonial structure be administered and interact with other space communities?

(g) **Frontier impact**. What will be the long-term impact of a nearly endless propagation of frontier settlements and communities? How will the presence of these frontier enclaves affect the general economic and social health of the more mature settled space communities?

(h) **Illegal activities**. Inevitably there will be those who will take advantage of others in the new wilderness of space. How will sanctions be administered? Under what authority? How will conflicting laws, rules, policies of nations, companies, and independent space communities be administered?

(i) **Settlement life cycle**. Will the birth/death cycle of outposts and communities follow similar patterns? If so, then what are the key indicators of impending growth or probable decline? How do these community life cycles compare with similar Earth communities?

(j) **Efficacy of phases of adoption model**. An examination of the expected progression of the five-phase model of adoption. How does the Everett Rogers model of diffusion of innovation (1962) hold up in an environment devoid of preexisting economic activity?

(k) **Automation**. What is the long-term effect of automation on human communities in space? To what degree will automation technologies developed for application in space be adopted on Earth?

(l) **Technology compatibility.** Will new common standards of practice be adopted for space-based manufacturing and commerce? Of corollary interest is if new common standards are adopted for space enterprises, then will these new standards disrupt Earth-based competitors, customers, and supply chains?

(m) **Non-space corporate engagement.** Which companies that are presently not associated with space will recognize the new space economy as an opportunity to expand their customer base and brand?

(n) **Lessons learned.** Given that we don't know what we don't know, what will we come to know in 2050, 2075, and 2100 about the creation and transformation of space economy?

Appendix 3: Self-Quiz About Space: How Do you Measure Up?

Take a few moments to consider the following general questions about the future of space settlement and the space economy. You'll notice that these questions ask about the social, economic, political, and legal setting for space. There are no technical questions about traveling faster than the speed of light, the impact of flooding Earth's gold market with space gold, or how moon ice will be converted to rocket fuel. This is about other dimensions of the new space economy.

There are no right or wrong answers so feel free to answer these questions about space as honestly as you can. Take the quiz and answer these same questions about assumptions and visions of space at the end of the book so you can measure the change from your earlier responses against your response after exploring more about space settlement in the new cosmos economy.

Space Quiz

1. Space development will take place at an orderly and law-abiding manner.   T or F
2. Human space settlement will be well established   _____
    (a) In 25 years
    (b) In 50 years
    (c) In 100 years
    (d) Never

3. When human space settlement is well established, Earth will remain as the central economic, political, social, and legal hub of human civilization.   T or F
4. When human space settlement is well established, China and the U.S. will remain as the central economic, political, social, and legal powers. T or F
5. Space settlement will likely fix Earth's social, political, and economic problems. T or F
6. The primary value of the space economy will be based upon acquisition and refining of raw materials as well as manufacturing, and energy production.   T or F
7. The space economy will be consumer-based. T or F
8. Space frontiers will be tamed in short order and replaced by mature markets.   T or F
9. The space economy will function independently from Earth's economy.   T or F
10. Space law will uniformly set standards for justice and sanctions.   T or F
11. The main reason to settle space is to escape the effects of climate change.   T or F
12. Only mega-corporations will benefit from settlement.   T or F
13. Space is the manifest destiny of humankind. T or F
14. Settlement will provide a haven for those who wish to escape civilization   T or F
15. Industrializing space will benefit Earth by moving toxic production off planet   T or F
16. The space economy and settlement will mostly benefit elites T or F
17. Human space settlement is necessary to alleviate Earth's overpopulation   T or F
18. The best use of space is for military purposes   T or F
19. I would like to visit the Moon or Mars as a space tourist   T or F
20. I would like to live permanently in space   T or F

Index

A
Adoption phase, 72, 237
Agri-business, 51, 138, 142, 181
Agricultural revolution, 129
Agriculture
   archaic, 138
   for feeding and commercial livestock, 139
   modern large-scale grain production, 181
   space-based, 137, 138, 141
Agriculture products, 129
Agri-robots, 141
Alternating current (AC), 187
American frontier, 34
American immigrants, 176
American Manifest Destiny, 34
American technology, 209
Animal products, 139
Anti-settlement proponents, 67
Artemis, 95, 124, 125
Artemis Accords, 202–204
Artificial intelligence (AI), 85
Assumptions, 62, 63
   aerospace companies, 61
   civilized commercial space, 62
   climate change, 66
   customers and consumers, 64
   emotional fervor, 65
   human activities, 61, 65
   limitation, 64
   PPP, 65
   rail access, 64
   space economy, 64
   space professionals, 62
   space settlement, 66
   species extinctions, 65
   technologic developments, 61
   workshops, 61
Asteroid mining, 135
Asteroids, 139, 167, 174
   aluminum, 151
   giant 3D printer, mining operation, 155
   metallic, 146
   mining operation, 155
   mining process, 146
   ore-bearing, 148
   Psyche 16, 150
   resources, 146
   robotic probes, 149
   space mining, 147
   transportation, 153
Automation, 162, 296
Autonomous economic zones, 257
Autonomous mining, 147

B

Balancing arguments, 97
Bezos, Jeff, 30, 111–113, 133, 154
Bigelow, 121, 122
Bigelow Expandable Activity Module (BEAM), 121, 168
Blue Ocean Strategy, (2005), 25
Blue Oceans
 business investment, 25
 citizens, 27
 development, 27
 humans, 31
 knowledge, 29
 revenue minus expenses, 26
 science fiction, 28
 social commitment, 27
 social responsibility, 27
 social skills, 31
 space frontier, 25
 space investor, 26
 space markets, 26
 space ventures, 29
Blue Origin, 109–112, 118, 280
Branson, Richard, 30, 37
British Commonwealth, 173
Budding business ideas, 284
Business and political sectors, 201
Business context, 211
Business model, 73, 223, 246, 278, 296
 business plan, 223
 commercial activity, 224
 common, 226
 economic adoption, 225
 examples, 224
 frontier space economy, 227
 investment, 227
 memory, 226
 parts, 224
 platforms, 227
 PPP model, 228
 and strategies, 227
 time and place, 223
 western frontier, 223
Business-to-business (B2B) commerce, 93, 273

C

Carbon-based fuels, 137
Celestial outpost, 140
Center of gravity (CG), 143
Challenge
 competitive rivalry, 98
 frontier justice, 99
 limited infrastructure, 99
 monopolistic competition, 101
 robots *vs.* people, 100
 space itself, 101
 start-up, 97, 98
 technology, 98, 99
China, 84, 85, 93, 127
Chinese lunar exploration program, 145
Christensen, Clayton, 74
Cislunar, 168–169
Classic economics, 179
CNBC, 125
Cobots, 155, 157, 158
 deployment, 164
 HRI, 164
 in diagnostic medical services, 164
 manufacturing tasks, 163
 medical, 164
 specialized, 163, 164
Collaborative commons, 85, 86
Collaborative consumption, 83
Collective national memory, 209
Colonialism, 296
Colonies, 84, 85, 107, 114, 160
Colony, 173, 174
Commerce, 136
Commercial chain, 89
Commercial investments, 134
Commercial space, 119–121
Commercial space economy, 97

Commercial space industry, 119
Commodities speculator, 278
Common refrain, 148
Communities, 83–87, 89, 108, 110
Competitive industries, 97
Competitive Strategy (1980, 2004), 70
Complete Adoption Model Synopsis, 259
Complex space economy, 284
Computer revolution, 89
Consumer market, 154
Corporate enterprises, 67
Cosmos economy, 39, 209, 236, 251, 258, 259, 284, 286, 289
   entrepreneurs, 283
   investment argument, 284
Covid19 pandemic, 156, 179
Creativity, 71

D

Direct current (DC), 187
Down-mass, 63, 90–92, 99
Dutch East India Company, 217

E

Early adopter phase, 221, 239, 258
Early mainstream adoption, 248
Early mainstream phase, 221, 249
   economic trading, 247
   growth and expansion, 246
   potential market adoption, 248
Earth-based projects, 210
Earth-bound social programs, 67
Earth consumer markets, 57
Earth's economic development, 61
Earth's global economy, 272, 292
Earth–Lunar axis, 136
Earth's market sectors, 99
Earth's surface, 48
Earth's traditional global economy, 293
Earthmen
   accountable tasks, 53
   constant stress, 54
   prison solitary confinement, 53
   social isolation, 53
   space, 53, 55
   submarine isolation, 53
Economic activities, 72, 75, 272
Economic adoption, 72
Economic and political power, 259
Economic and social adoption, 259
Economic force, 118
Econosphere, 169
Edison, Thomas A., 187, 193
Emergency assistance, 203
Emerging market phase, 249
Emerging markets, 269
Emotional intelligence (EQ), 31, 241, 267
Energy, 136, 137, 141
   and food production in space, 148
   space environment, 154
Energy generation, 137, 196
Entrepreneur, 26, 28, 89, 90, 97, 99, 108, 112, 283
Entrepreneurial capital investments, 175
Entrepreneurial firms, 120
Entrepreneurial space organization, 211
European Space Agency (ESA) members, 190
Everett Rogers model, 296
Exchange traded funds (ETFs), 134

F

Fair and just legal system, 99
Financial transactions, 295
Forego space investment, 210
Frankenaut, 57
Free-floating hothouse plantations, 142
Frontier, 35, 36, 39, 81, 82, 87, 89, 90, 99, 105, 108, 217, 218, 257, 259, 260

Frontier impact, 296
Frontier-level economic centers, 248
Frontiers, 33–37
 knowledge, 34
 optimism, 33
 poverty and persecution, 34
 solar system, 37
 space, 38
 U.S. western, 37

G

Gamma rays, 279
Gateway, 168–170
Global parochialism, 284
Global positioning system (GPS), 120
Gold, 147–149, 151
Greenfield market, 25
Greenhouses, 142
Gross space product (GSP), 143

H

Hostile environment, 290
Human-centric space mission, 51
Human factors, 136
Human physical enhancement, 58
Human space settlement, 209
Human vs. automated space activities, 288
Human/Robot interaction (HRI), 164
Hybrid system, 85

I

Illegal activities, 296
Imitative growth, 236
Immigrant migrations, 171
Immigrants, 141, 170, 171, 179, 180, 183, 184
In situ resource utilization (ISRU), 168, 195

Indian Space Research Organization (ISRO), 189
Industrial manufacturing, 148, 150, 151
Industrial revolution, 128–130, 150
Industrial space, 180, 269, 277
 creativity, 215–217
 frontier and colonial commerce, 217–219
 innovation, 215–217
 managing space, 219, 220
 phase of development, 220, 221
Industrial/B2B market, 154
Industrialized solar system, 257, 279
Infrastructure, 93, 97, 99, 112, 113, 122, 149, 182
 groundwork, 171
 space communities, 175
Infrastructure and transportation capabilities, 111
Infrastructure investment, 161
Initial public offering (IPO), 120
Innovation, 71, 239
Innovative business approach, 73
Innovative initiatives, 71
Innovators, 243
In-situ resource utilization (ISRU), 141, 242
In-space supply chain, 120
Intellectual property (IP), 180
International Space Station (ISS), 121, 125
International Traffic in Arms Regulations (ITAR), 205, 206
Internet of things (IoT), 164
Interoperability, 203
Investment, 133, 134, 149
Investors, 89
ispace (a private Japanese company), 145
Israel, 205

J
Japan, 134–136, 189
Japan's Aerospace Exploration Agency (JAXA), 189

K
Kahneman, Daniel, 243
Keiretsu-like partnerships, 245
Kennedy, John F., 81

L
Lagging Adopter phase, 258
Lagging adopters, 254
Large-scale commercial businesses, 34
Late mainstream phase, 220, 252
    mainstream adopters, 250
    penultimate phase signals, 250
Leadership, 20, 23
Leading business schools, 215
Learning, 98
Lewin, Kurt, 216
Long tail business model, 233
    entrepreneurs, 234
Low Earth orbit (LEO), 120, 249
Luddites, 66
Lunar, 95, 120, 125, 140
Lunar Orbital Platform, 169

M
Machiavelli, Niccolo, 69
Manned space enterprise, 162
Manufacturing, 141
    BEAM modules, 168
    cobots, manufacturing tasks, 163
    electric light business, 180
Manufacturing, space
    efficiency, 157
    fiber-optics cables, 154
    heavy industrial, 154
    micro-gravity, 156
    OEMs, 154
    raw material extraction, 153
    space environment, 154
    technical limits, 154
    teleoperation, 155
    3D bio-manufacturing, health-related products, 154
    transportation, 153
    ZBLAN fiber optic cable, 156
Manufacturing, space pharmaceuticals, 154
Manufacturing, space solar technology, 154
Market disruption, 73
Market leaders, 98
Market maturity phase
    growing space economy, 252
    stage, 252
Markets, 90
Mars, 54, 95, 101, 104, 113, 114, 117, 123, 125, 135, 136, 139, 140, 159–161, 167, 169, 189
    communications, 149
    harvesting water, 148
    robotic exploration, 156
    self-identify, 183
    tele-robotics, 156
Maslow, Abraham, 44
Medical cobots, 164
Mergers and acquisitions (M&A), 194
Micro-gravity environment, 54
Micro-gravity manufacturing, 135
Microscopic bug-like animals, 56
Military, 98, 103
Mining, 90, 113, 119, 122, 125, 127, 135, 136
    asteroid operation, 155, 156
    asteroids, 146–148
    global mining operations, 145
    gold, 147–149, 151
    harvesting mineral riches, 149
    metallic asteroid, 146
    and refining, 149
    robots use, 163

Mining water, 139
Modern strategic planning process, 70
Moon, 82, 100, 108, 125, 135, 139, 140, 145, 146, 182, 188, 190, 202, 209
  autonomous mining, 147
  cislunar space, 169
  harvesting water, 148
  International Space Station, 169
  landings, 160
Multi-phase process, 31, 257
Multi-system approach, 85
Musk, Elon, 107, 110, 113, 114, 122, 189, 241

N

NASA, 53, 54, 94, 118, 125, 128, 135, 140, 147, 160, 168, 169, 203, 204, 293
National space agencies, 135
National Space Council, 48
Natural satellites, 136
New space economy, 277, 283
New space merchants
  capitalist value chain, 127
  central planning, 127
  commercial space, 128
  firms, 128
  growth, 128
  industrial revolutions, 128–130
  non-space companies, 128
  private sector, 127
  space investments, 127
  space nations, 127
Non-disclosure agreement (NDA), 266
Non-space corporate engagement, 297
Non-space firms, 182
Northrup Grumman (NGC), 121
Noxious production facility, 150

O

O'Neill, G.K., 160–162, 167
Open source business model, 232, 233
Opportunity costs, 97
Optimal business model, 198
Original equipment manufacturers (OEMs), 154
Outposts, 84, 85, 94, 97, 100, 101, 111, 120, 135, 136, 148, 182
Outposts, space
  cislunar, 169
  colony, 173
  deep space station, 174
  definition, 174
  goods and services, 176
  interspace, 169
  place in space, 172
  space immigrants, 172
  station, 174
  trading, 174

P

Personal computers (PCs), 73
Petroleum energy supply, 147
Pharmaceuticals, 154
Phase adoption model
  adopters, 239
  characteristics, 239
  commercial activities, 245
  commercial locations, 243
  early adopters, 244
  enterprises, 237
  formative frontier stage, 242
  frontier economy, 245
  global economy, 237
  innovator phase, 239
  lagging adapter phase, 239
  market adoption, 244
  marketplace, 239
  micro economies, 237
  new space economy, 239

organizations and countries, 242
phase description, 238
phases, 239
prototypes, 237
rate, 239
S-Curve path, 238
social and economic agenda, 241
soft skills, 241
space-based commercial
    customers, 246
space-based market economy, 237
space economic development, 238
Pilgrims
    careers, space, 265, 266
    competition, 264
    earth-bound counterparts, 263
    emotional investment, 265
    financial opportunity, 263
    greenfield opportunities, 263
    innovative leader, 266, 267
    intrinsic motivators, 264
    lack of social skills, 264
    planet, 263
    profit and power, 263
    quality/value, 264
    rehabilitation, 264
    space business, 268
    space community, 264
    space jobs, 268
    trade and commerce, 265
Plan Dragon, 113, 114
Planning, 266
Platinum, 145
Playing field, 211
Poets and quants, 4
Pollution, 150
Potential economic autonomy, 250
Potential market adoption, 250
Principle of Cumulative
    Advantage, 49, 98
Private space investment, 67
Public–private partnership (PPP), 65,
    228, 242, 279, 291
    benefits, 228, 230
    economic impact, 229
    public sector, 228
    railroads, 229
    SOE, 230
    in the U.S, 230
Private space companies, 136

R

Railroad business model, 229
Redwire, 123
Resource-based economy, 85, 86
Return on investment (ROI), 26, 217
Risk-averse investors, 90
Robotic mining operations, 149
Robotic probes, 149
Robotic spacecraft technologies, 145
Robots, 32, 155, 172, 272, 280
    benefits, 162
    and cobots, 157, 163
    disposable and expendable, 162
    exploration, 156
    HRI, 164
    industrial equipment
        companies, 163
    industrial technologies, 163
    labor-intensive responsibilities, 155
    and other IoT, 164
    programmed, 161
    programmers, 162
    self-managed, 163
    space economy, 163
    space exploration, 162
    telerobotic control, 156
    tele-robotics, 156
Rogers, Everett, 237
Rules of the game
    foreboding land of regulations,
        205, 206
    industrial space, 201
    policy, 201–204
    settlements, 206, 207
    space economy, 201
    space law, 201–204

S

Science fiction, 28
Sci-Fi books, 91
S-Curve model, 197, 198
Self-contained settlement outpost, 155
Settlement adoption, 142
Settlement communities
  collaborative commons, 85, 86
  communes, 86, 87
  communities, 86, 87
  community capitalism, 84, 85
  consumerism, 82–84
  consumption, 82–84
  frontier, 81
  legislative initiatives, 82
  New Frontier programs, 81
  resource-based economy, 85, 86
  social programs, 82
  solar system, 82
  space settlement, 84
Settlement communities social systems, 84
Settlement life cycle, 296
Settlement, space, 167
  cislunar, 169
  commercial space, 181
  community, 167, 174
  development, 180
  diversity, 182
  Earth Age, 184
  goods and services, 176
  humanist movement, 181
  in human history, 183
  ISRU construction, 168
  networked space, 182
  place in space, 172
  private firms, 176
  space community formats, 172
  tight-knit settlement community, 179
Settlements, 188, 193, 195, 197, 207, 263, 265, 271
Seven Habits of Highly Effective People, 70
Skittish investors, 97
Social and economic stability, 259
Social diversity, 108
Social issues, 129
Social value, 129
Solar panel business, 30
Solar system, 69, 89, 125, 207
  Earth's economy, 195–197
  financial elements, 193
  role of competition, 193–195
  S-Curve, 197–199
  service providers, 199
  space communities, 193
  space innovation, 195–197
  supply chains, 199
Southern slaveholders, 17
Space, 18, 19, 279, 280, 282, 289, 290, 296
  benefits, 187, 188
  businesses operating, 50
  colonists, 160
  commercial activities, 142
  commercial competition, 188
  competition, 188
  critical government communications, 188
  destiny, 18, 19
  economic risks and opportunities, 21
  economy, 19, 134, 148
  emerging business, 22
  experiment, 140
  exploration, 140
  food production, 137, 138
  global commercial interest, 22
  globalization 2050, 189–191
  industrialization, 133
  infrastructure, 133, 137, 141
  inhabitants, 153
  investment, 133–137
  leadership, 188
  mankind, 18
  manufacturing (*see* Manufacturing, space)

mining (*see* Mining)
myths and superstitions, 24
national/corporate platforms, 187
Plan Alpha, 93
Plan Beta, 93
planet's environment, 154
planning projections, 20
primarily industrial, 153
science fiction, 18, 28
sector, 19
self-interest, 187
settlement, 136, 139, 141, 143, 151, 159, 160, 162
solar system, 93, 94, 188
space-centric market, 154
stigma, 17
tourism, 134
US/NASA dominance, 21
vision, 154
vital infrastructure, 163
Space Act of 2015, 146
Space adventurers, 47
Space agriculture, 137
Space Angels, 133
Space angels network, 122
Space-based agri-business, 138, 142
Space-based agriculture industry, 141
Space-based business, 292–293
Space-based capital ventures, 210
Space-based capitalism, 284
Space-based commerce, 280
Space-based economic engine, 257
Space-based economy, 20, 92, 142, 242
Space-based markets, 48
Space-based technologies, 274
Space business investment, 39
Space businesses, 143
Space commerce, 63
Space economy, 37, 38, 47, 49, 72, 74, 75, 98, 201, 270–273, 275, 276, 278, 281, 289
    aerospace and defense industrial complex, 118
    commercial space, 117–121
    cosmos economy, 117
    data analytics/consumer communications, 117
    exit to profitability, 124, 125
    global economic development opportunities, 117
    new space investors, 121–123
    private space companies, 118, 119
    technical and financial capabilities, 117
    traditional business sectors, 117
Space enterprise, 89, 117
Space entrepreneurs, 97
Space environment, 101
Space exploration, 43
    citizens and business Investors, 44
    climate change, 44
    frontier, 46
    human emigration, 43
    pull/pus, 45
    space settlement, 43, 46
Space Force, 43
Space Foundation, 128
Space frontier, 46, 47
Space immigrants, 170–172
Space industry, 268, 275
Space investment, 67, 182
Space policies, 201
Space robots, 168
Space settlement, 66, 210, 278
Space settlement economy, 97, 143, 210
Space settlement movement, 46
Space settlers, 139, 140
Space survivability, 57
Space Tourism, 243
Space traveler's environment, 56
Space Treaty, 202
Space-centric ETFs, 134
Spacefarer, 55
SpaceFund, 119, 123, 133, 136
SpaceFund Reality (SFR) rating, 136

Space-generated energy, 137
Spaceman, 56
Spaceports, 91, 92
Space-related transport technology, 30
SpaceX, 109, 110, 118, 121, 123, 268, 280
SPCE (Virgin Galactic stock symbol), 133, 134
Standard operating procedure (SOP), 127
State owned enterprises (SOE), 230, 242
  in China, 230
  organization, 230
Stations
  cislunar location, 174
  deep space, 174
  International Space Station, 168, 169
  place in space, 172
  ULA, 169
Steep barrier, 98
Strategic thinking, 69
Successful business ventures, 72
Supply chain, 136, 274, 295
Sustainable settlement economy, 141

T

Talent, 275
Technology compatibility, 297
Telemedicine, 156
Telemetry, 156
Teleoperation, 149, 155, 156
Telerobotic control, 156
Tele-robotics, 156
Tesla, Nikola, 187, 189
Theory of Human Motivation (1943, 44
Time, 281
Tourism, 108, 110, 118, 121, 125, 127
Toxic industrial activities, 150
Trade, 136
Trading outposts, 136
Traditional bottom line, 181
Traditional business enterprises, 235
Traditional business models, 253
Traditional business plan, 278
Traditional business sectors, 117
Traditional exchange model, 235
Traditional retail business model, 224
Transcontinental railroads, 64
Transparency, 203
Trans-planetary economy, 283
Transportation, 136
Triple bottom line, 181
Turner Thesis (1893), 33
Turner, Frederick Jackson, 108

U

United Arab Emirates (UAE), 189
Universal basic income (UBC), 85
U.S. Commerce Department, 22

V

Venture capital (VC) firms, 119, 133
Venturesome, 240
Vertical integration, 231, 232
Vertical integration business model, 232
Virgin Galactic (VG), 133, 134
Virginia Company, 217
Virtual reality (VR), 156
Visions, 170
  O'Neill's, 160
  of man's future in space, 160
  of networked space settlements, 182
  space utopia, 160
Visions of space
  arguments, 104
  commercial ventures, 104
  constructive and positive frame, 106
  cosmic neighborhood, 105
  destiny, 105
  Earth, 103, 104
  economic gain, 105

economic opportunity, 105
economic pragmatism and expansion, 108
food and potable water, 105
humans, 103
Laissez-faire approach, 104
military/political strategic high ground, 107
mission statement matchup, 109, 110
Plan Blue, 111, 112
Plan Dragon, 111, 113, 114
Plan Z, 114, 115
Plan-B strategy, 106
Planet chauvinists, 112, 113
planet Earth, 106
science fiction adventures, 105
sense of control, 104
social and political experimentation, 108
social diversity, 108
space settlement, 107
supporting human settlements, 104
theological/spiritual rationalization, 108
variance of thinking, 103
western frontier/modern space settlers, 105
Volatile economic setting, 211
Von Clausewitz, 69
Vonnegut, Kurt, 4

W

Water, 94
West Side stockyards, 34

Z

ZBLAN fiber optic cable, 156

About the Author

Jack Gregg has designed, launched, and managed top-ranked MBA programs at four prominent universities and has consulted with public and private organizations on strategic professional development and organizational change. He is currently an adjunct professor of management. Dr. Gregg is a long-time resident of Southern California where he lives with his wife and best friend, Chihiro, along with their cats, Tora and Stinky.

GPSR Compliance

The European Union's (EU) General Product Safety Regulation (GPSR) is a set of rules that requires consumer products to be safe and our obligations to ensure this.

If you have any concerns about our products, you can contact us on

ProductSafety@springernature.com

In case Publisher is established outside the EU, the EU authorized representative is:

Springer Nature Customer Service Center GmbH
Europaplatz 3
69115 Heidelberg, Germany

www.ingramcontent.com/pod-product-compliance
Lightning Source LLC
LaVergne TN
LVHW010336260326
834688LV00036B/731